AF360816

Bridging the Gap: From Biomechanics and Functional Morphology of Plants to Biomimetic Developments

Bridging the Gap: From Biomechanics and Functional Morphology of Plants to Biomimetic Developments

Editors

Olga Speck
Thomas Speck

MDPI • Basel • Beijing • Wuhan • Barcelona • Belgrade • Manchester • Tokyo • Cluj • Tianjin

Editors
Olga Speck
Cluster of Excellence livMatS
@ Freiburg Center for
Interactive Materials and
Bioinspired Technologies
(FIT)
University of Freiburg
Freiburg, Germany

Thomas Speck
Plant Biomechanics Group @
Botanic Garden Freiburg
University of Freiburg
Freiburg, Germany

Editorial Office
MDPI
St. Alban-Anlage 66
4052 Basel, Switzerland

This is a reprint of articles from the Special Issue published online in the open access journal *Biomimetics* (ISSN 2313-7673) (available at: https://www.mdpi.com/journal/biomimetics/special_issues/plants_bomimetic).

For citation purposes, cite each article independently as indicated on the article page online and as indicated below:

LastName, A.A.; LastName, B.B.; LastName, C.C. Article Title. *Journal Name* **Year**, *Volume Number*, Page Range.

ISBN 978-3-0365-8140-8 (Hbk)
ISBN 978-3-0365-8141-5 (PDF)

Contents

About the Editors

Olga Speck

Olga Speck studied biology and sports at the University of Freiburg (Germany) and received her PhD in 2003 on vibration damping in plants. Since 2002, she supervised research and development projects within her research interests: (i) the functional morphology and biomechanics of plants; (ii) bioinspired materials systems such as self-repairing materials systems, adaptive materials systems and composite materials; (iii) biomimetics and sustainable technology development; and (iv) education and training in the fields of biomechanics and biomimetics. She is Principal Investigator in the Cluster of Excellence "Living, Adaptive and Energy-autonomous Materials Systems (livMatS)" at the University of Freiburg. At the interface between Area C "Longevity" and Area D "Societal Challenges", she supervises research projects on damage control in plants such as self-repair and abscission as a model for technical applications and sustainable technology solutions.

Thomas Speck

Thomas Speck studied biology at the University of Freiburg (PhD 1990) and received the Venia Legendi for Botany and Biophysics in 1996. After a visiting professorship at the University of Vienna, he was offered professorships at the Humboldt-Universität zu Berlin and at the University of Freiburg, where he was Associate Professor for "Botany" and the Director of the Botanic Garden from 2002 until 2006. After declining the offer of a full professorship and directorship of the Botanic Garden at the Freie Universität Berlin, he became a Full Professor for "Botany: Functional Morphology and Biomimetics" in Freiburg in 2006. He is a spokesperson of the Cluster of Excellence "Living, Adaptive and Energy-autonomous Materials Systems (livMatS)" and of the Competence Network Biomimetics, and is Vice-Chair of the Society for Technical Biology and Bionics. Thomas Speck is Deputy Managing Director of the Freiburg Center for Interactive Materials and Bioinspired Technologies (FIT) and a scientific member of the Materials Research Center Freiburg (FMF).

Preface to "Bridging the Gap: From Biomechanics and Functional Morphology of Plants to Biomimetic Developments"

The Special Issue "Bridging the Gap: From Biomechanics and Functional Morphology of Plants to Biomimetic Developments" collects seven articles. One article concerns the challenges and potential pitfalls of the biomimetic approach. The remaining articles cover the entire development chain, from basic research in the field of biomechanics and the functional morphology of plants, to simulations and the development of physical models for a better understanding of functional principles, ultimately leading to biomimetic products at the laboratory scale or demonstrator level.

We thank all the authors for their valuable and thoughtful contributions and the reviewers for their helpful comments to improve the manuscripts.

Olga Speck and Thomas Speck
Editors

 *biomimetics*

Editorial

Bridging the Gap: From Biomechanics and Functional Morphology of Plants to Biomimetic Developments

Olga Speck [1,2,*] and Thomas Speck [1,2]

1 Plant Biomechanics Group @ Botanic Garden Freiburg, University of Freiburg, D-79104 Freiburg, Germany; thomas.speck@biologie.uni-freiburg.de
2 Cluster of Excellence *liv*MatS @ FIT—Freiburg Center for Interactive Materials and Bioinspired Technologies, University of Freiburg, D-79110 Freiburg, Germany
* Correspondence: olga.speck@biologie.uni-freiburg.de; Tel.: +49-761-203-2803

Citation: Speck, O.; Speck, T. Bridging the Gap: From Biomechanics and Functional Morphology of Plants to Biomimetic Developments. *Biomimetics* **2021**, *6*, 60. https://doi.org/10.3390/biomimetics6040060

Received: 15 October 2021
Accepted: 15 October 2021
Published: 18 October 2021

Publisher's Note: MDPI stays neutral with regard to jurisdictional claims in published maps and institutional affiliations.

1. Introduction

During the last few decades, biomimetics has attracted increasing attention in both basic and applied research and in various fields of industry and building construction. Biomimetics has a high innovation potential and opens up possibilities for the development of innovative technical products and production chains. The specific structures and functions that the vast number of organisms have evolved in adaptation to differing environments represent the basis for all biomimetic R&D projects. Novel sophisticated methods for quantitative analyses and simulations of the form–structure–function relationship at various hierarchical levels provide intriguing insights into multiscale mechanics and other functions of biological materials and surfaces. For the first time, it is possible to transfer biological structures and thus their properties into innovative biomimetic products by means of newly developed production methods and at reasonable cost.

Animals, with their fascinating behaviour and movement processes, have long attracted interest, with plants only more recently having also been recognised as valuable concept generators for biomimetic research [1,2]. In general, from a material scientist's point of view, plants can be regarded as fibre-reinforced materials systems defined by a number of material properties [3]. These plant materials systems consist of various components with different material properties and, thus, are not only anatomically inhomogeneous and mechanically anisotropic, but also possess a spatial and temporal heterogeneity attributable to growth and their capacity to respond or adapt to changing environmental conditions.

2. Broad Spectrum of Topics

The articles published in this Special Issue, "Bridging the Gap: From Biomechanics and Functional Morphology of Plants to Biomimetic Developments", cover theoretical considerations concerning the challenges of the biomimetic approach and potential pitfalls and include the entire development chain from basic research in the field of biomechanics and functional morphology of plants to simulations and the development of physical models for a better understanding of functional principles, finally leading to biomimetic products on the laboratory scale or demonstrator level. The size scale includes all hierarchical levels from microstructure to the entire plants.

Niklas and Walker [4] detail the considerable challenges of the biomimetic approach, which uses approaches to model and abstract the behaviour or properties of living systems based on inference within structure–function relationships for the development of synthetic bioinspired structures and systems. The authors discuss potential pitfalls by comparing the ways in which engineering and biological systems are analysed operationally, address the challenges of modelling biological systems, and suggest some methods for assessing the validity of these models.

In the second contribution, Wunnenberg et al. [5] describe the strengthening structures in the petiole-lamina transition zone of peltate leaves, in which the petiole is centrally attached to the lamina. In such peltate leaves, which are found in 357 plant species, the transition from the rod-like petiole to plate-like lamina is characterised by a marked change in geometry. The authors have analysed the connection between petiole and lamina in 41 species, with a particular focus on the reinforcing fibre strands. They have discerned several design principles that can be used as models for plant-inspired lightweight supporting structures.

Masselter et al. [6] present models for 3D reticulated actuator systems inspired by the macroscopic cortical fibre networks found in some extant and fossil plants. The asymmetric deformation of these networks caused by asymmetric secondary wood growth enable the up-righting of inclined balsa and papaya stems. This functional movement principle has been transferred to elastic technical hollow tubes that are surrounded by a net-like structure. In addition, the influence of fibre angles on deformation behaviour under internal pressure is analysed and described.

Klang and Nickel [7] describe the development of biomimetic freeze-casted graded ceramics inspired by the spines of the lance sea urchin. These spines possess a plant-like hierarchical lightweight construction and represent superstructures with several gradation features including porosity, pore orientation and pore size. The spines have considerable biomimetic potential for porous ceramics with predetermined breaking points and adaptable behaviour under compression up to failure. Some of these features can be included in an abstracted way in ceramics manufactured by freeze-casting.

Poppinga et al. [8] show that a variety of complex plant movement principles can be demonstrated with comparably simple handcrafted compliant systems based on paper, wood, plastic foil and/or glue as construction materials. The handcrafted systems are self-actuated by shrinking processes triggered by the anisotropic hygroscopic properties of the wood or paper. The developed systems have a high potential for fast, precise and low-cost abstraction and transfer processes in biomimetics and for the "hands-on understanding" of plant movements in university and school courses.

Mühlich et al. [9] have analysed, through simulation and physical testing, the influence of several design criteria, such as stiffness and hinge width, in compliant folding mechanisms moved by bioinspired pneumatically actuated hinges composed of fibre-reinforced plastic. The authors have developed a workflow within a finite element model software that allows mathematical models to be inferred for the prediction of mechanical properties and of deformation behaviour as a function of the parameters used as design criteria.

Krüger et al. [10] have developed a material design space for 4D-printed bioinspired bilayer structures with hygroscopic actuation that takes into account unequal effective layer widths and deflections under self-weight. The curvature of various bilayer strips has been described using an adapted Timoshenko model, and its ability to predict curvatures has been validated in experiments. This approach has led to an analytical solution space enabling the quantification of the influence of Young's moduli, swelling strains and densities on deflection under self-weight and curvature under hygroscopic swelling. In addition, it allows the selection of a suitable material combination in bioinspired bilayer systems with unequal layer widths.

The articles published in this Special Issue, "Bridging the Gap: From Biomechanics and Functional Morphology of Plants to Biomimetic Developments", thus, give an up-to-date overview of current research topics in plant-inspired biomimetic research.

Author Contributions: Both authors contributed equally to this editorial. All authors have read and agreed to the published version of the manuscript.

Funding: The work of the guest editors is funded by the Deutsche Forschungsgemeinschaft (DFG, German Research Foundation) under Germany's Excellence Strategy—EXC-2193/1—390951807.

Acknowledgments: As guest editors, we extend our thanks to all authors who contributed to this Special Issue and to the reviewers who provided helpful comments that have improved the papers in this Special Issue.

Conflicts of Interest: The authors declare no conflict of interest.

References

1. Speck, T.; Speck, O. Quo vadis plant biomechanics–Old wine in new bottles or an up-and-coming field of modern plant science? *Am. J. Bot.* **2019**, *106*, 1–5. [CrossRef] [PubMed]
2. Speck, O.; Speck, T. Functional morphology of plants—A key to biomimetic applications. *New Phytol.* **2021**, *231*, 950–956. [CrossRef] [PubMed]
3. Wegst, U.G.K.; Bai, H.; Saiz, E.; Tomsia, A.P.; Ritchie, R.O. Bioinspired structural materials. *Nat. Mat.* **2015**, *14*, 23–36. [CrossRef] [PubMed]
4. Niklas, K.J.; Walker, I.D. The challenges of inferring organic function from structure and its emulation in biomechanics and biomimetics. *Biomimetics* **2021**, *6*, 21. [CrossRef] [PubMed]
5. Wunnenberg, J.; Rjosk, A.; Neinhuis, C.; Lautenschläger, T. Strengthening structures in the petiole–lamina junction of peltate leaves. *Biomimetics* **2021**, *6*, 25. [CrossRef] [PubMed]
6. Masselter, T.; Speck, O.; Speck, T. 3D reticulated actuator inspired by plant up-righting movement through a cortical fiber network. *Biomimetics* **2021**, *6*, 33. [CrossRef] [PubMed]
7. Klang, K.; Nickel, K.G. The plant-like structure of lance sea urchin spines as biomimetic concept generator for freeze-casted structural graded ceramics. *Biomimetics* **2021**, *6*, 36. [CrossRef] [PubMed]
8. Poppinga, S.; Schenck, P.; Speck, O.; Speck, T.; Bruchmann, B.; Masselter, T. Self-actuated paper and wood models: Low-cost handcrafted biomimetic compliant systems for research and teaching. *Biomimetics* **2021**, *6*, 42. [CrossRef] [PubMed]
9. Mühlich, M.; González, E.A.; Born, L.; Körner, A.; Schwill, L.; Gresser, G.T.; Knippers, J. Deformation behavior of elastomer-glass fiber-reinforced plastics in dependence of pneumatic actuation. *Biomimetics* **2021**, *6*, 43. [CrossRef] [PubMed]
10. Krüger, F.; Thierer, R.; Tahouni, Y.; Sachse, R.; Wood, D.; Menges, A.; Bischoff, M.; Rühe, J. Development of a material design space for 4D-printed bio-inspired hygroscopically actuated bilayer structures with unequal effective layer widths. *Biomimetics* **2021**, *6*, 58. [CrossRef]

 *biomimetics*

Article

Development of a Material Design Space for 4D-Printed Bio-Inspired Hygroscopically Actuated Bilayer Structures with Unequal Effective Layer Widths

Friederike Krüger [1,*,†], Rebecca Thierer [2,3,*,†], Yasaman Tahouni [3,4], Renate Sachse [5], Dylan Wood [3,4], Achim Menges [3,4], Manfred Bischoff [2,3] and Jürgen Rühe [1]

[1] Laboratory for Chemistry and Physics of Interfaces, Department of Microsystems Engineering, University of Freiburg, Georges-Koehler-Allee 103, 79110 Freiburg, Germany; ruehe@imtek.uni-freiburg.de
[2] Institute for Structural Mechanics, University of Stuttgart, Pfaffenwaldring 7, 70550 Stuttgart, Germany; bischoff@ibb.uni-stuttgart.de
[3] Cluster of Excellence Integrative Computational Design and Construction for Architecture (IntCDC), University of Stuttgart, Keplerstraße 11, 70174 Stuttgart, Germany; yasaman.tahouni@icd.uni-stuttgart.de (Y.T.); dylan.wood@icd.uni-stuttgart.de (D.W.); achim.menges@icd.uni-stuttgart.de (A.M.)
[4] Institute for Computational Design and Construction (ICD), University of Stuttgart, Keplerstraße 11, 70174 Stuttgart, Germany
[5] Institute for Computational Mechanics, School of Engineering and Design, Technical University of Munich, Boltzmannstraße 15, 85748 Garching b. München, Germany; renate.sachse@tum.de
*   Correspondence: f.krueger.bionik@gmail.com (F.K.); thierer@ibb.uni-stuttgart.de (R.T.)
†   These authors contributed equally to this work.

Citation: Krüger, F.; Thierer, R.; Tahouni, Y.; Sachse, R.; Wood, D.; Menges, A.; Bischoff, M.; Rühe, J. Development of a Material Design Space for 4D-Printed Bio-Inspired Hygroscopically Actuated Bilayer Structures with Unequal Effective Layer Widths. *Biomimetics* **2021**, *6*, 58. https://doi.org/10.3390/biomimetics6040058

Academic Editor: Jose Luis Chiara

Received: 22 July 2021
Accepted: 30 September 2021
Published: 6 October 2021

Abstract: (1) Significance of geometry for bio-inspired hygroscopically actuated bilayer structures is well studied and can be used to fine-tune curvatures in many existent material systems. We developed a material design space to find new material combinations that takes into account unequal effective widths of the layers, as commonly used in fused filament fabrication, and deflections under self-weight. (2) For this purpose, we adapted Timoshenko's model for the curvature of bilayer strips and used an established hygromorphic 4D-printed bilayer system to validate its ability to predict curvatures in various experiments. (3) The combination of curvature evaluation with simple, linear beam deflection calculations leads to an analytical solution space to study influences of Young's moduli, swelling strains and densities on deflection under self-weight and curvature under hygroscopic swelling. It shows that the choice of the ratio of Young's moduli can be crucial for achieving a solution that is stable against production errors. (4) Under the assumption of linear material behavior, the presented development of a material design space allows selection or design of a suited material combination for application-specific, bio-inspired bilayer systems with unequal layer widths.

Keywords: material development; biomimetic bilayer actuators; hygroscopic actuation; 4D-printing; mechanical modeling

1. Introduction

Hygroscopically actuated movements can be observed in various plant structures. Their function reaches from protection of pollen [1] to active seed dispersal and seed burial [2–4] to seed protection [5,6]. Often, these movements are caused by functional bilayer mechanisms made from the combination of a restrictive passive layer and a responsive active layer, where cellulose fibers control the direction of expansion as a response to tissue hydration [5,7,8]. These swelling effects occur without the need for metabolic energy consumption and can therefore also be observed in dead tissues [6]. These attributes have raised the interest of similarly functioning biomimetic bilayer systems in application fields such as sensor technology [9], medicine [10], and architecture [11–13]. Deforming bilayer structures have been built from a variety of materials such as hydrogel [14,15], wood veneer [11,12], or fiber composites [16,17].

More recently, 4D-printing has been used to develop structures with programmable hygroscopic movement. Using the fused filament fabrication (FFF) method, bending motion analogous to that of the biological role models can be achieved by an anisotropic arrangement of swelling fibers of a commercial wood-filled filament [18–20].

For most of these material systems, the influence of basic geometric changes such as layer thickness are well understood, enabling computational reproduction of observed curvatures based on Timoshenko's model for bimetal bending, adapted for hygromorphic materials [12,16,17,21–23]. This knowledge allows for designing new demonstrators with planned and timed curving behavior [18]. However, if active and passive layer effective widths differ, e.g., due to variable spacing between adjacent printing paths or the width of the 3D-printed strands, Timoshenko's formula has to be modified. Furthermore, to our knowledge, the influence of material parameters on curvature has not yet been analyzed systematically. Such information enables a more application-oriented material selection or development. Fields in which bilayer systems have to be designed for more than the resulting curvature would especially benefit, e.g., as in architectural applications. Here, not only the curvature of a structure due to swelling or shrinking has to be considered for a successful application, but also its deflection under self-weight, absorbed water, or wind, depending on the arrangement and support conditions of the system. Easily applicable models and measurements are favorable, because complicated modeling and simulation and extensive experimental parameter measurements are time and material consuming. To focus on the application of bio-inspired bilayer structures that aim to perform repetitive hygroscopically actuated motions, the choice of an elastic material is crucial. Therefore, Timoshenko's simple model [23] is chosen as a starting point.

In this paper, Timoshenko's model [23] is adapted to account for unequal effective widths of the active and passive layer. The model is validated by means of an established hygromorphic, 4D-printed bilayer system, where swelling strains and Young's moduli of the used materials are obtained by simple, single-material experiments. For validation, various experiments on bilayers are conducted and compared to the computed values. Based on the adapted model and on basic linear bending theory, the interplay of various material parameters on curvature due to changes in moisture content and on deflection due to self-weight, including absorbed water, will further be analyzed. This is then used for the development of a material design space with fixed geometric parameters and variable values for Young's moduli, swelling strains, and self-weights.

2. Materials and Methods

2.1. Development of Mechanical Models

We used a modeling approach for predicting the curvature due to changes in moisture content of bilayer cantilevers with unequal effective layer widths that establishes an extension of Timoshenko's model for composite beams [23]. Considering the thicknesses t_a, t_p and Young's moduli E_a, E_p of the active and passive layer, respectively, and the total thickness $h = t_a + t_p$ of the beam, Timoshenko's equation provides an analytical solution for the curvature radius r and the curvature κ caused by a swelling strain ε_{swell} in the longitudinal direction in the active layer.

To account for unequal effective widths of the active and passive layer, caused by the porous mesostructure of the 3D-printed layers, we incorporated the effective layer widths b_a, b_p into the equation. The effective layer widths b_a and b_p were obtained by subtracting the width of gaps g between adjacent printing strands in the longitudinal direction from the total width of the respective layer and thus only considering the materialized part of the total width (Figure 1a). The analytical solution for the curvature then reads:

$$\kappa = \frac{1}{r} = \frac{6\varepsilon_{swell}(1+m)^2}{h\left(3(1+m)^2 + (1+nmq)\left(m^2 + \frac{1}{nmq}\right)\right)} \text{, with } n = \frac{E_p}{E_a}, \ m = \frac{t_p}{t_a}, \text{ and } q = \frac{b_p}{b_a}. \tag{1}$$

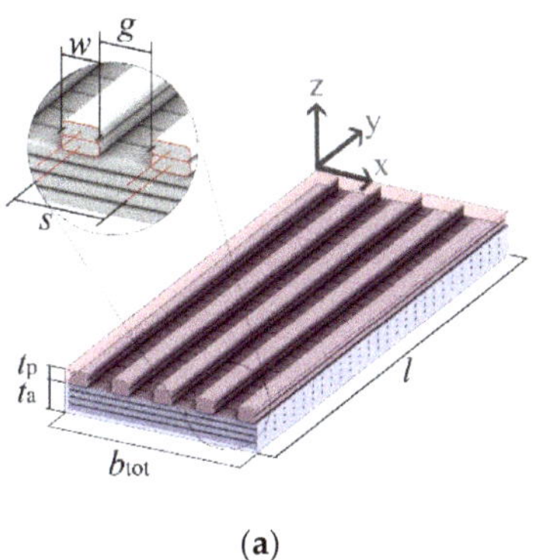

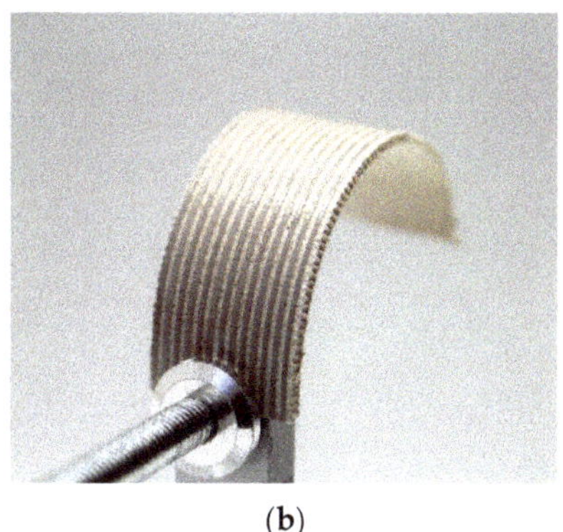

(**a**)   (**b**)

Figure 1. Geometry of bilayer samples. (**a**) Schematic of bilayer sample consisting of active (blue) and passive (red) layer. l: Length of bilayer. t_p: Thickness of passive layer. t_a: Thickness of active layer. w: Width of printed strands (extrusion width). g: Gap between adjacent strands. s: Spacing between adjacent printing paths. Effective width of active layer $b_a = b_{tot}$. Effective width of passive layer was calculated as $b_p = b_{tot} - n_{gap} \cdot g$, with number of gaps n_{gap}. Filament deposition along the x-axis was defined as transversal, along the y-axis as longitudinal. Single-material layers of the same geometry were used as samples for measurements of Young's moduli. (**b**) Photograph of bilayer sample in dry and therefore curved state. Grey strands of passive material are visible.

A complete derivation of the formula can be found in Appendix A. Referring to laminate theory, this model resembles the classical laminate theory (CLT), where material fibers normal to the mid-surface remain straight, normal and unstretched during deformation. Transverse shear effects are therefore excluded. The formulation can handle large rotations, but is restricted to small strains.

For the deflection of the cantilevers under self-weight, including absorbed water, Bernoulli's geometrically linear static bending theory of beams was used. The cross-sectional porosity of the bilayers was taken into account by using effective layer widths to calculate the ideal bending stiffness of composite beams.

2.2. Measurements of Established System's Parameters

2.2.1. Sample Production

Laywood meta 5 filament (LayFilaments, Cologne, Germany) from two different batches was used as active layer material and combined with a generic PLA filament (SUNLU, Zhuhai, China) as passive layer material to print multi-material bilayer structures via fused filament fabrication (FFF). Both materials were printed on a dual-extruder FFF 3D-printer (FELIX TEC 4, FELIXprinters, IJsselstein, The Netherlands) equipped with two 0.5 mm brass nozzles and a heated bed. A custom toolpath design and G-code generation workflow, which was built in Rhinoceros 3D CAD environment and Grasshopper Plugin (McNeel and Associates), was used to generate digital designs and G-codes (Voxel2GCode) for 3D-printing [18].

In all samples, the printing parameters, including nozzle temperature, bed temperature, E-value (volume of extruded filament per mm), material layer height, and feed rate was kept constant (Table A1). By fixing these parameters, it was assured that the width of a single 3D-printed strand w was kept constant. The effective layer width b could then be controlled by the toolpath spacing parameter s, which is defined as the distance between the two adjacent printing toolpaths (Figure 1a) and controls the width of the gap g between two adjacent strands. In all active layers, 3D-printed strands were oriented transversally and in passive layers longitudinally. In all samples, active layers were printed with a spacing of 0.5 mm, resulting in a solid layer. Passive layers were printed with a spacing of 1.5 mm (except those of experiment C). Samples with different thicknesses in the active (experiment A) and passive (experiment B) layer were manufactured by varying the number of printed material layers while keeping a constant material layer height of 0.2 mm. In experiment A the number of active material layers was chosen as 2, 4, 6, 7 and 8. In experiment B the number of passive material layers was chosen as 1, 2, 4, 6 and 8. To analyze samples with different effective passive layer width (experiment C), spacing between passive layer strands was chosen as 0.5 mm, 0.6 mm, 1.0 mm, 1.5 mm and 3 mm. For each experiment, seven samples were printed and tested.

Following the manufacturer's instructions, samples with Laywood meta 5 were rinsed for five days after printing and dried at 30% humidity for another two days to reach their responsive state. After this treatment, a repeatable response to water submerging and drying was assumed, and all samples were successively investigated in wet, humid, and dry state. For that, samples were placed in a climate chamber (KBF 115, Fa. BINDER GmbH, Tuttlingen, Germany) at 23 °C in water submersion for wet, in 80% ambient humidity for humid and in 30% ambient humidity for dry state over two days to reach equilibrium.

2.2.2. Measurements of Young's Modulus and Swelling Strains

According to Equation (1), Young's moduli of active and passive layers E_a and E_p and the swelling strain ε_{swell} have to be determined for the curved humid and dry states. Additionally, for the deflection due to self-weight, Young's moduli in wet state are required.

Young's moduli were determined from deflections of eight monolayer samples for Laywood meta 5, tested repeatedly in all three states, and eight monolayer samples of PLA for each state. Laywood meta 5 samples were 75 mm × 20 mm × 0.8 mm with transverse filament deposition and PLA samples were 75 mm × 20 mm × 0.2 mm with longitudinal filament deposition, identical to the printing direction in the bilayer structures (Figure 1a). Length, width, thickness and weight of these samples were measured in all states to calculate the swelling strain in all directions and self-weight in all states, using a digital caliper with an accuracy of 0.02 mm and a scale with an accuracy of 0.001 mg.

2.2.3. Bilayer Deflection and Curvature Measurements

Deflection under self-weight and curvature upon humidity change were measured on 3D-printed bilayer samples (Figure 2). Deflection under self-weight was measured in wet state. Samples were clamped at their base for 3 mm and images were taken perpendicular to their edge (Figure 2a). Images were then analyzed with ImageJ [24,25] and deflection was measured as the vertical distance between the clamped surface and the tip of the beam. To compensate for initially slightly curved sample geometries, deflections were measured with passive layers facing upward as well as downward and their mean values were taken.

Curvatures were measured in humid and in dry state. Samples were positioned on a flat surface, standing parallel to their strong axis to avoid additional bending due to self-weight (Figure 2b). Images were taken from above and analyzed with a script in ImageJ, calculating the radius r of the circumference of a circle defined by three points, selected on the sample edge [24–26].

After measuring deflection and curvature, all bilayer samples were again submerged in water, their layers deconstructed and geometry changes of resulting monolayers were again measured in wet, humid, and dry state.

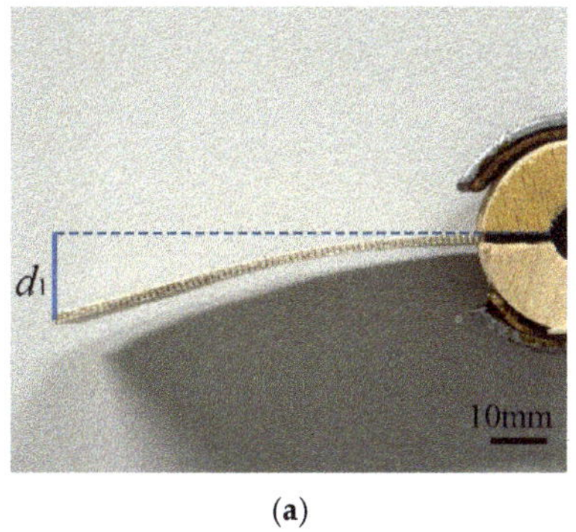
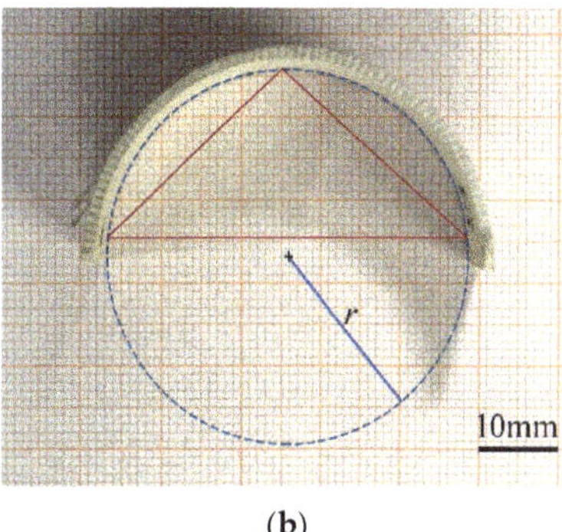

(a)   (b)

Figure 2. Evaluation of deflection and curvature. (**a**) Deflected sample in wet state. Measurement of deflection as the vertical line between sample base and tip; (**b**) curved bilayer sample in dry state. Evaluation of curvature by selecting three points on the sample, getting the defined triangle and calculating the radius of its circumcircle.

3. Results

3.1. Material Parameters of the Established Bilayer System

Values for Young's moduli are very different between Laywood and PLA with the ratio of Young's moduli n in the range of 25 and 60 for dry and humid state, respectively (Table 1). For Laywood, an approximately 10-fold increase from wet to dry state can be observed, while the Young's moduli for PLA show a small increase from wet to dry state.

Table 1. Young's moduli in all states. Measurements are given as median (IQR). Values were recalculated from deflections under self-weight. $N = 8$.

	Young's Modulus in MPa		
	Wet	**Humid**	**Dry**
Laywood, batch 1	6.3 (0.8)	37.6 (11.4)	99.0 (12.8)
Laywood, batch 2	9.7 (0.5)	61.9 (10.5)	110 (36.9)
PLA	1785 (558)	2189 (303)	2467 (569)

Generally, no major differences between lengths in the second and third rinsing cycle could be found in monolayer samples of both Laywood batches (Figure 3). An exception was the length of wet samples of batch 2, where median lengths of 75.5 mm in the second and 75.2 mm in the third rinsing cycle were observed. As the following bilayer measurements were taken in the second rinsing cycle, these values were taken for further calculations. Length changes for all samples were calculated, defining the wet state as 100%, leading to a shrinkage of $\varepsilon_{swell} = 2.6\%$ from wet to humid and 3.5% from wet to dry state for batch 1 (2.8% and 3.7% for batch 2, respectively). For PLA no shrinkage could be observed (Figure 3).

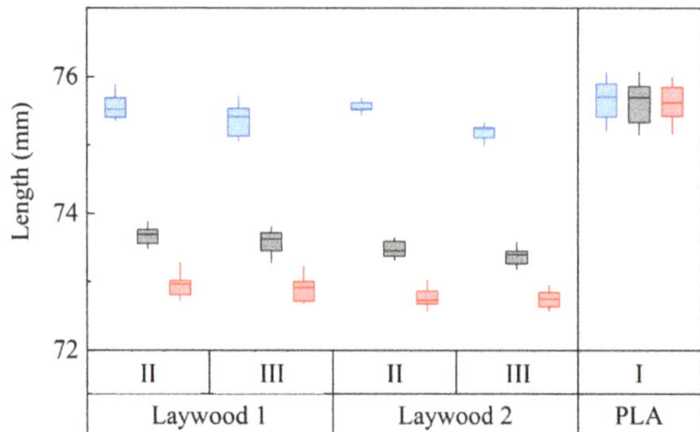

Figure 3. Length development of repeatedly tested samples of Laywood meta 5 and PLA. Blue: wet state. Gray: humid state (80% RH). Red: dry state (30% RH). Roman numerals refer to rinsing cycles. Laywood 1 and 2 refer to different batches of material. No differences can be seen between the second and third rinsing cycles in Laywood. Only Laywood samples show a major length change between humidity levels. $N = 8$.

3.2. Comparison of Measured Bilayer Deflections and Curvatures with Computed Values

Deflections under self-weight were calculated in wet state, including the absorbed water, according to Bernoulli's geometrically linear static bending theory of beams. All curvatures due to hygroscopic swelling were modeled for the transition from wet to humid and wet to dry state, using the modification of Timoshenko's formulation (Equation (1)).

Deflections and curvatures were calculated for every tested sample of experiments A (varying active layer thickness), B (varying passive layer thickness) and C (varying passive layer effective width), using individual geometry input data from respective measurements in wet, humid, or dry state. For the active material, we used median Young's moduli and swelling strains of Laywood batch 1 for experiment A and median values of Laywood batch 2 for experiments B and C. For the passive material of all samples, we used median Young's moduli of PLA, respectively. Calculated and measured deflections and curvatures for every sample are compared and analytical curves for a hypothetical, median-based cantilever are used to broaden the range of the respective free variable (Figure 4). This analytical solution for the deflections and curvatures is given for a hypothetical cantilever, which is defined by median values of the measured geometric and material parameters (Tables A2 and A3). For all computed values, there is no major difference between using sample-wise measured geometric input values or using the geometry of the hypothetical, median-based cantilevers. The computed deflections are all in good accordance with the measured values (Figure 4a–c). Although the absolute values differ slightly, the trends of a decreasing curvature for increasing layer thicknesses are clearly visible for both measured and calculated values in experiments A and B (Figure 4d,e,g,h). In experiment A, the measured values show a high variation, especially for thin active layers (Figure 4a,d,g). For experiment C, an opposing trend is predicted by the curvature calculation than can be seen in the measured curvature values (Figure 4f,i).

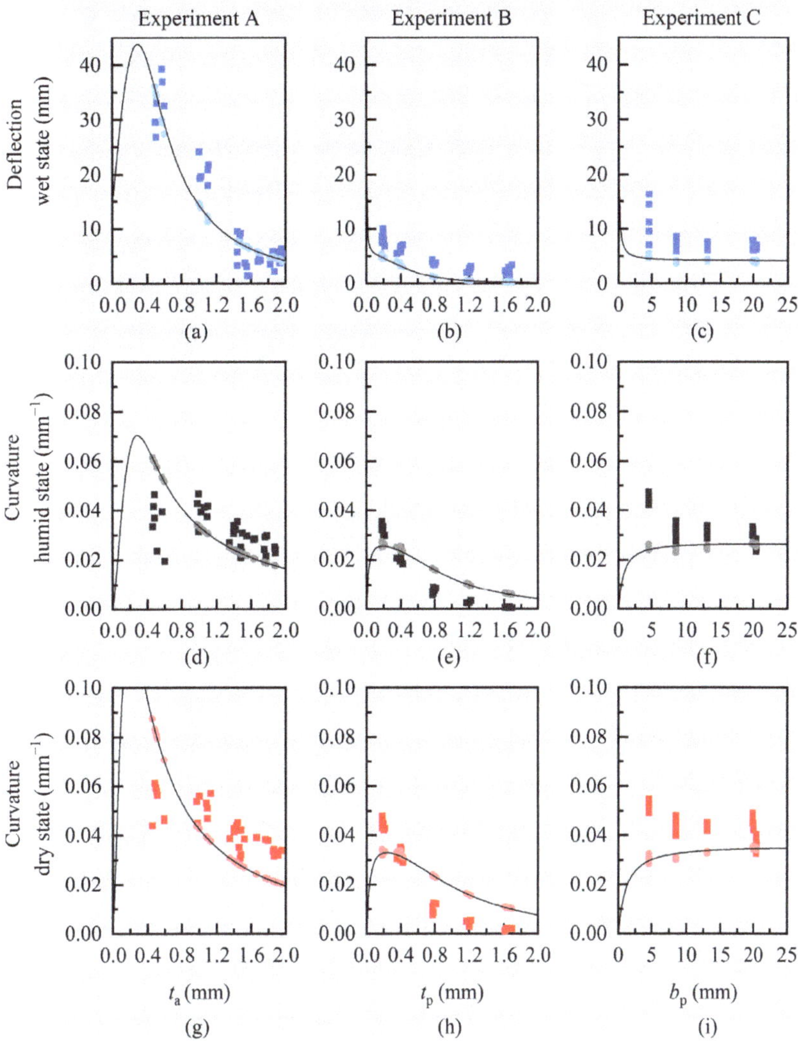

Figure 4. Measured and computed values for deflection and curvature of Laywood/PLA bilayers. (**a–c**) Blue: Deflection in wet state; (**d–f**) Grey: Curvature values in humid state; (**g–i**) Red: Curvature values in dry state. Dark squares: measured values. Light circles: computed values using the analytical models and geometric input values as measured for every individual sample. Lines represent the solution of the analytical models for a hypothetical bilayer cantilever with median values of measured geometries. All computed deflections are in good accordance with the measured values. For increasing thicknesses of active and passive layers, the computed curvatures show similar trends to the measured values with slight differences in absolute numbers. For increasing passive layer effective width, the computed and measured curvatures show opposing trends.

4. Discussion

4.1. Evaluation of Computed Deflections and Curvatures

As can be seen in Figure 4, the overall trends for most of the computed deflections and curvatures match the experimental data. However, for experiment C with modification of the effective width of the passive layer b_p the analytical model predicts an opposite trend for the curvature. Additionally, all values show slight variations between measured and calculated results. To examine the potential of the modified model in Equation (1) to predict the experimentally measured curvature, we fit the swelling strain ε_swell and the ratio of Young's moduli n to the measured curvature values of the experiments by least squares approximation. Thus, we were able to separate effects from strongly scattering material parameters from our validation of the structural model. Here, we used again the hypothetical, median-based cantilever geometries together with the medians of measured curvatures (Table 2).

Table 2. Results for least squares fit of material parameters n and ε_swell to measured curvatures in humid and in dry state for original and modified Timoshenko model.

Experiment		Modified Model (Equation (1))		Original Model (Timoshenko)	
		n	ε_swell	n	ε_swell
Humid state	A	691.2	0.034	286.0	0.034
	B,C	610.4	0.043	149.2	0.037
Dry state	A	561.9	0.045	233.9	0.045
	B,C	299.7	0.051	111.6	0.048

As the samples of experiment A were produced from a different batch of Laywood material than in experiments B and C, we performed separate optimizations (Table 2). A fit of parameters for Timoshenko's original model, which assumes the equality of layer widths for the curvature calculation, was done for comparison (Table 2). With these resulting material parameters, both models are able to depict the measurements of experiments A and B, while only the modified version can reproduce the measured decreasing trend of curvature in experiment C with increasing passive layer effective width b_p (Figure 5).

The resulting elongations ε_swell from the optimization approach (Table 2) are slightly larger than those measured in single-material experiments. These differences can be explained by inaccuracies of length measurements in wet state. The extremely soft material might have been compressed during the manual measurements, resulting in an underestimation of wet sample length. The resulting ratios of Young's moduli n from the optimization approach differ strongly from those measured in all single-material tests, where we found ratios of about $n = 40$. This could be due to a violation of the assumption of linear material behavior, which is a major element of Timoshenko's model [23]. Tensile tests showed the end of the elastic range at approximately 2% for Laywood in humid and in dry state (Supplementary Materials File S1). Additionally, in single-material deflection tests of humid and dry state, a maximum elastic strain of only approximately 0.1% could be calculated (Supplementary Materials File S2), while for curved bilayers a maximum elastic strain of about 5% resulted from calculations using the modified model, which means an exceedance of the elastic limit (Supplementary Materials Files S3 and S4).

Assuming non-linear material behavior and therefore redefining n as the ratio of stiffnesses between passive and active material, changes between this ratio from $n = 40$ at elastic strains of 0.1% to $n = 600$ at strains of 5% could occur, if the materials showed softening effects for high strains. A strong hint towards plastic deformations in the active layer can be found when reviewing the length development of deconstructed bilayers: If deformations would only occur in the elastic range, lengths of deconstructed Laywood monolayers should be the same as in single layers of Laywood. This expected behavior can be seen for deconstructed Laywood layers from experiments A and C (Figure 6). However, in experiment B a strong increase in length could be observed, suggesting that plastic deformations occur when the passive layer is relatively thick and the bilayers experience rather a low curvature but high elastic strains in the active layer (Supplementary Materials File S4). Such plastic deformations should be avoided, e.g., by choosing a different material for the active layer, if the bilayer structure is intended to perform repetitive hygroscopically actuated motions.

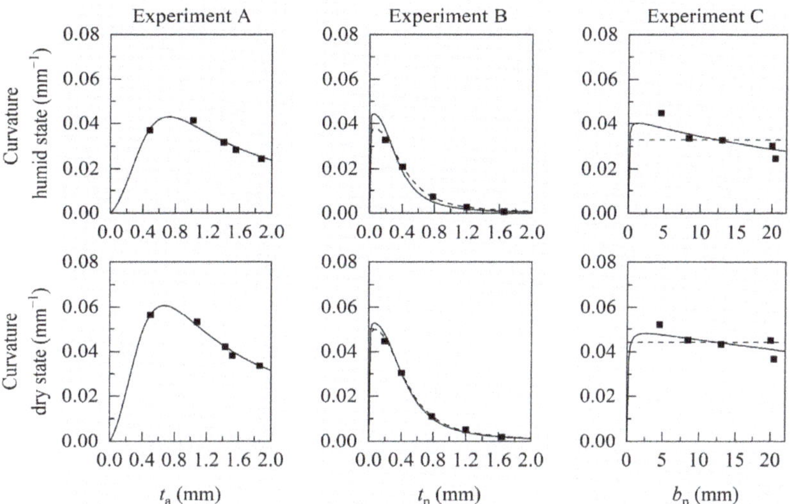

Figure 5. Analytical curves with optimized ratios of Young's moduli and swelling strains. Squares: Median values from experiments. Solid line: Using optimized parameters for the modified model. Dashed line: Using optimized parameters for Timoshenko's original model. For Experiment A with different active layer thickness t_a, lines from optimizations are congruent. Both models are able to depict the measured trends of experiments A and B, while only the modified model is able to reproduce a decreasing trend of curvature for increasing passive layer effective width b_p in experiment C.

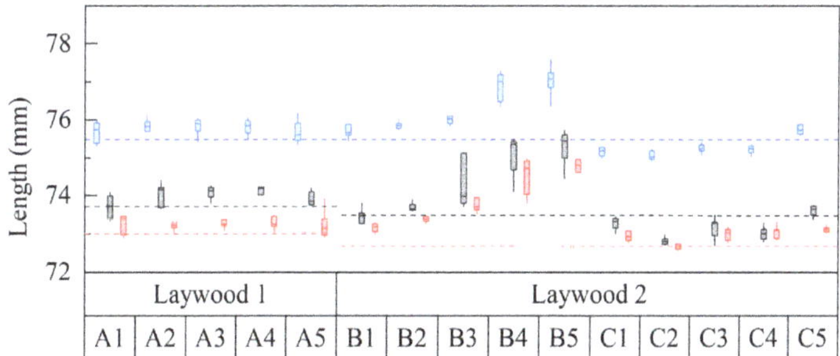

Figure 6. Length development of Laywood meta 5 monolayers from deconstructed bilayers. Blue: wet state; black: humid state; red: dry state; A1 to A5: increasing active layer thickness; B1 to B5: increasing passive layer thickness; C1 to C5: decreasing passive layer width. Laywood 1 and 2 refer to different batches of used Laywood meta 5 filament. Dashed lines mark the median length of pure Laywood meta 5 monolayers. With increasing passive layer thickness, a distinct increase in active layer length is visible. $N = 7$.

Although the observed experimental values could not be reproduced computationally, because the potentially non-linear material behavior is not accounted for in the model, we could show that by extending the original model of Timoshenko by the effect of individual effective layer widths, a tool for the curvature prediction of bilayer cantilevers with unequal effective widths, e.g., due to 4D-printing, is provided. We therefore use the modified Timoshenko model in the following to build the material design space.

4.2. Development of a Design Space

To allow suitable material selection, we consider the relation between curvatures of bilayers due to humidity changes and their deflections under self-weight. Large curvatures in dry state are accompanied by large deflections in wet state. Combinations of thin active and passive layers lead to an increasing curvature, but also result in very large deflections (Figure 4a,g). An increase in passive layer effective width decreases deflection, but inhibits bending as well (Figure 4b,h). Consequently, if only geometric parameters of a bilayer system can be changed, a trade-off between large curvature and limited deflection has to be made with respect to the desired use of the bilayer structures. A good starting point for this trade-off can be found by using the analytical formulas for curvature (Equation (1)) and deflection to define a design space. To do so, all geometric parameters have to be set.

For an exemplary design space, we set the geometric parameters as measured for samples of experiment A in dry and in wet state with a median active layer thickness of $t_a = 1.43$ mm in dry and $t_a = 1.45$ mm in wet state (Figure 7). The density of the passive material is chosen as measured with $\rho_p = 1.43$ mg/mm^3 and the passive layer's Young's moduli $E_p = 1785$ MPa (lower surface in Figure 7b) and $E_p = 1000$ MPa (upper surface in Figure 7b) are considered.

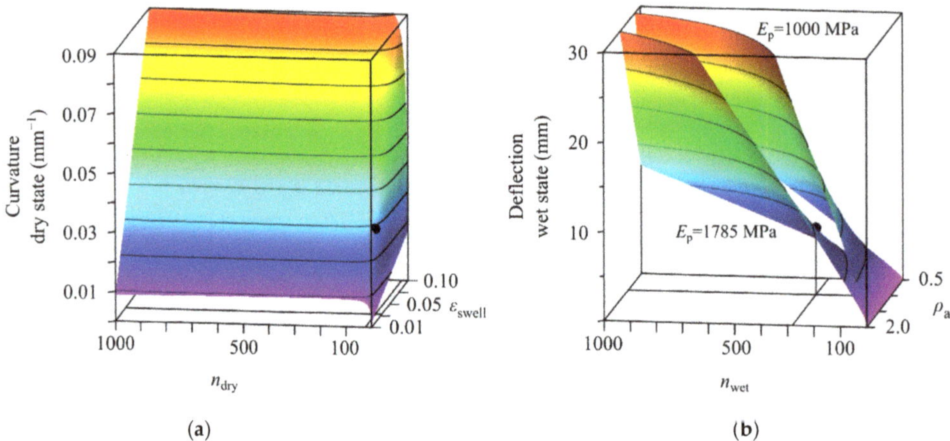

Figure 7. Exemplary design space for the choice of parameters of an active material for a given geometry and different assumptions on the passive material. (**a**) Solution surface of curvature for various swelling strains ε_{swell} and ratios of Young's moduli n. Colors indicate magnitude of curvature. (**b**) Solution surfaces of deflection for various densities ρ_a and ratios of Young's moduli n. The lower surface corresponds to an absolute value of the passive layer's Young's modulus $E_p = 1785$ MPa, the upper surface corresponds to $E_p = 1000$ MPa. For both surfaces the density of the passive material is chosen as $\rho_p = 1.43$ mg/mm^3. Colors indicate magnitude of deflection. Black dots mark the solutions for the given Laywood/PLA bilayers of samples from experiment A with $t_a = 1.43$ mm in dry and $t_a = 1.45$ mm in wet state.

The solution surface for the curvature space shows a distinct kink at around $n_{dry} = 30$. Below this ratio, small changes of n_{dry} lead to drastic changes in the resulting curvature. The swelling strain, however, has only a linear influence on the curvature. When analyzing deflections, the active layer's density ρ_a and n_{wet} have an almost linear influence on the deflection. It can be lowered by decreasing either ρ_a or n_{wet}. Another option would be the simultaneous increase of E_p and E_a.

4.3. Usage and Limitations of Material Design Space for Material Selection or Development

After formulating the analytical solutions for the individual material design space, one target value each for curvature and deflection can be chosen. We recommend using the target curvature to choose values for n_{dry} and swelling strains. Here, attention has to be drawn to existing kinks in the solution surfaces, as can be seen in the exemplary solution in Figure 7a. The values should not be chosen too close-by to achieve a stable prediction, if the material parameters used later for the actual bilayer system show typical scattering due to production errors. If commercially available or already developed materials are used, the corresponding value n_{wet}, which will differ from n_{dry}, if material's stiffnesses are moisture dependent, can be calculated. If not, we recommend an educated guess to determine the range of n_{wet}. With the set value of n_{wet}, absolute values for E_a and E_p can be chosen from the analytical solution for deflections, leading to a resulting density of the active material. With these values at hand, it is possible to reach the target values for curvature and deflection. Additionally, further fine-tuning or variation of curvature and deflection can be realized by reiterating on the geometric parameters, such as layer thicknesses and spacing. This can also be useful to program the timing of the curvature process [18], which is so far not included in the presented mechanical model and material design space.

After this step, additional considerations are necessary. First, the justification of all important assumptions has to be checked. As could be seen, non-linear material behavior at the maximum strain resulting from curvature or deflection, can lead to differences between measured and calculated values. A more accurate modeling approach including FEM simulations would be needed (Appendix D). However, such an approach lacks an analytically closed solution and cannot be easily used to build a material design space as proposed. Additionally, if the chosen material has a high density or low stiffness in the curved state, additional deflections have to be considered in a combined load case of self-weight and hygroscopic swelling.

Other limitations can be system inherent, e.g., minimum material layer heights or the stepwise material layer height increase resulting from 3D-printing. Another effect is delamination events, which are pronounced in the presented system for passive layer thicknesses beyond 1.2 mm and passive layer widths below 4.5 mm.

Possible other limitations and the applicability of the proposed material design space will have to be investigated in future work. Another field of further research could include the timing of the deformation process into the mechanical model.

In conclusion, the curvature of the mesoporous printed bilayers can be predicted by a slightly modified Timoshenko model, which uses the effective widths of the active and passive layers. A proper choice of a suitable material combination for the generation of actuatable bilayer structures can be accelerated by developing a specific design space that is based on these simple analytical formulas. Especially the influence of small differences in the material parameters on the resulting geometry can be identified and taken into account when developing new materials.

Supplementary Materials: The following are available online at https://www.mdpi.com/2313-7673/6/4/58/s1, File S1: Tensile tests; File S2: Values Laywood monolayers; File S3: Bilayers strain distribution maximum; File S4: Bilayers strain distribution thick bilayers.

Author Contributions: Conceptualization, F.K.; Formal analysis, R.T.; Funding acquisition, A.M., M.B. and J.R.; Investigation, F.K. and R.T.; Methodology, F.K., R.T. and Y.T.; Supervision, R.S., D.W., A.M., M.B. and J.R.; Visualization, F.K., R.T. and Y.T.; Writing—original draft, F.K. and R.T.; Writing—review and editing, Y.T., R.S., D.W., M.B. and J.R. All authors have read and agreed to the published version of the manuscript.

Funding: We acknowledge funding by the Ministry of Science, Research and the Arts Baden-Wuerttemberg for funding within the collaborative project "Bio-inspirierte Materialsysteme und Verbundkomponenten für nachhaltiges Bauen im 21tenJahrhundert" (BioElast)—MWK 33-7533-30-121/15/3, which is part of the Zukunftsoffensive IV "Innovation und Exzellenz—Aufbau und Stärkung der Forschungsinfrastruktur im Bereich der Mikro- und Nanotechnologie sowie der neuen Materialien". We also acknowledge the support by the Deutsche Forschungsgemeinschaft (DFG, German Research Foundation) under Germany's Excellence Strategy—EXC 2120/1—390831618. The article processing charge was funded by the Ministry of Science, Research and the Arts Baden-Wuerttemberg and the University of Freiburg in the funding program Open Access Publishing.

Institutional Review Board Statement: Not applicable.

Informed Consent Statement: Not applicable.

Data Availability Statement: The data presented in this study are available in Supplementary Materials.

Conflicts of Interest: The authors declare no conflict of interest. The funders had no role in the design of the study; in the collection, analyses, or interpretation of data; in the writing of the manuscript, or in the decision to publish the results.

Appendix A

Following Timoshenko's analysis of bi-metal thermostats [23], we consider a bilayer strip made of one active layer that can expand due to hygroscopic swelling in the longitudinal direction and one passive layer, which will not expand due to hygroscopic changes. If the bilayer experiences changing humidity conditions and the layers are ideally bonded interlaminarly, the passive layer prevents the strip from pure expansion, leading to bending of the strip. Assuming constant swelling and constant geometric parameters along the strip, the resulting curvature κ will be constant. Therefore, only an infinitesimal cross-sectional element is considered. Its geometric parameters are the thicknesses and widths of the active layer (t_a, b_a) and passive layer (t_p, b_p) and the total thickness of the strip $h = t_a + t_p$. The material parameters are the Young's moduli of both layers (E_a, E_p) and the longitudinal swelling strain of the active layer ε_{swell}.

In case of a positive swelling elongation ε_{swell} in the active layer, prevention of the passive layer from pure expansion leads to an axial compressive force P_a in the active layer. As we assume no additional external force loads, static equilibrium states a tensile force P_p in the passive layer, equal in amount: $P_a = P_p = P$.

The resulting sum of the bending moments in both layers are in equilibrium with the axial forces:

$$M_a + M_p = \frac{Ph}{2} . \tag{A1}$$

Assuming linear elastic material, the bending moments are related to the curvature via the bending stiffnesses:

$$M_a = E_a I_a \, \kappa, M_p = E_p I_p \, \kappa , \tag{A2}$$

with the respective second moment of area $I = \frac{bt^3}{12}$.

The assumption that fibers normal to the mid-surface, i.e., in the thickness direction, remain straight, normal and unstretched during deformation (Bernoulli's beam theory), leads to a kinematic conclusion, where the total strains at the interface of both layers must be equal:

$$\varepsilon_{a,tot} = \varepsilon_{p,tot} .$$

For small strains, the total strains of the active layer are additively decomposed into an elastic and a hygroscopic swelling part. The passive layer only experiences elastic strains:

$$\varepsilon_{a,elast} + \varepsilon_{swell} = \varepsilon_{p,elast} .$$

The elastic strains can be further split into a membrane and a bending part:

$$\frac{P_a}{E_a b_a t_a} + \frac{\kappa\, t_a}{2} + \varepsilon_{swell} = -\frac{P_p}{E_p b_p t_p} - \frac{\kappa\, t_p}{2}\,. \tag{A3}$$

Introducing the ratios $n = \frac{E_p}{E_a}$, $m = \frac{t_p}{t_a}$, $q = \frac{b_p}{b_a}$ and combining Equations (A1), (A2) and (A3) leads to Equation (1), i.e., the solution for curvature, valid for large rotations and small strains:

$$\kappa = \frac{1}{r} = \frac{6\varepsilon_{swell}\left(1 + m\right)^2}{h\left(3(1 + m)^2 + (1 + nmq)\left(m^2 + \frac{1}{nmq}\right)\right)}\,.$$

Appendix B

Table A1. Fixed printing parameters for all samples.

	Active Layer Laywood Meta 5	Passive Layer PLA
Layer height	0.2 mm	0.2 mm
E-value	0.040	0.033
Nozzle temperature	210 °C	210 °C
Bed temperature	55 °C	55 °C
Printing speed (feed rate)	1200 mm/min	1200 mm/min

Appendix C

Values of hypothetical, median-based cantilevers in wet (Table A2), humid, and dry state (Table A3).

Table A2. Hypothetical, median-based cantilever geometry and material input data for deflection calculations of bilayer experiments A, B and C in wet state.

	Experiment	t_a mm	t_p mm	b_a mm	b_p mm	L mm	ρ_a mg/mm^3	ρ_p mg/mm^3	E_a MPa	E_p MPa
	A	free	0.15	20.61	8.45	72.94	1.03	1.43	6.3	1785.4
wet state	B	1.45	free	20.69	9.10	73.11	1.02	1.03	9.7	1785.4
	C	1.46	0.22	20.69	free	72.64	1.02	0.99	9.7	1785.4

Table A3. Hypothetical, median-based cantilever geometry and material input data for curvature calculations of bilayer experiments A, B and C, in humid and in dry state.

	Experiment	t_a in mm	t_p in mm	b_a in mm	b_p in mm	E_a in MPa	E_p in MPa	ε_{swell}
	A	free	0.15	20.42	8.45	37.6	2188.7	0.026
humid state	B	1.38	free	20.43	9.10	61.9	2188.7	0.028
	C	1.37	0.22	20.48	free	61.9	2188.7	0.028
	A	free	0.15	20.30	8.45	99.0	2467.9	0.035
dry state	B	1.36	free	20.36	9.10	110.4	2467.9	0.037
	C	1.37	0.22	20.37	free	110.4	2467.9	0.037

Appendix D

Another modeling approach refers to a layer-wise theory (LWT) in finite strains. In contrast to CLT, displacements over the thickness are considered piecewise linear only within the various layers, i.e., the total displacement over the thickness can be C^0-continuous. Additionally, transverse shear effects are considered layer-wise. We performed simulations using the commercial finite element software ANSYS (Release 20 R1, ANSYS Inc. Canonsburg, PA, USA). We modelled the bilayer cantilever beams using two stacked layers of eight-node structural solid shell finite elements (SOLSH190) that are capable of handling large rotations and large strains and we used geometrically non-linear static analysis with a load controlled path following algorithm. Convergence studies showed that a mesh of 100 elements in the longitudinal direction and 30 elements in the direction of the cantilever width are needed per layer. To account for the porous mesostructure of the 3D-printed layers, Young's modulus of the passive layer was scaled, using the ratio of effective layer widths. The cantilever beam was clamped at one end and we applied a temperature load case, while defining the thermal expansion coefficient of the active layer in the longitudinal direction as the swelling strain ε_{swell}. For post-processing, we defined three nodes in the longitudinal direction, at both ends and in the middle of the cantilever, all located on the mid-surface. From the displacements of these nodes, we calculated the curvature radius and the curvature of the resulting arc. For the material, we choose a linear elastic definition on the one hand and a hyperelastic Neo-Hookean material law on the other hand, using the measured Young's moduli and a Poisson's ratio of $\nu = 0.0$.

The results for this modeling approach are exemplarily shown in Figure A1 for the bending of the median-based geometries of cantilevers from experiment B from wet to humid state (Table A3). Even for the relatively high thicknesses of active and passive layers ($t_a = 1.38$ mm), the differences in curvature results between the modified Timoshenko model, which corresponds to CLT, and the layer-wise modeling and simulation approach using a linear elastic material law are small. Consideration of a more realistic hyperelastic material law shows differing results for the highly curved samples that experience compressive elastic strains (Supplementary Materials File S3).

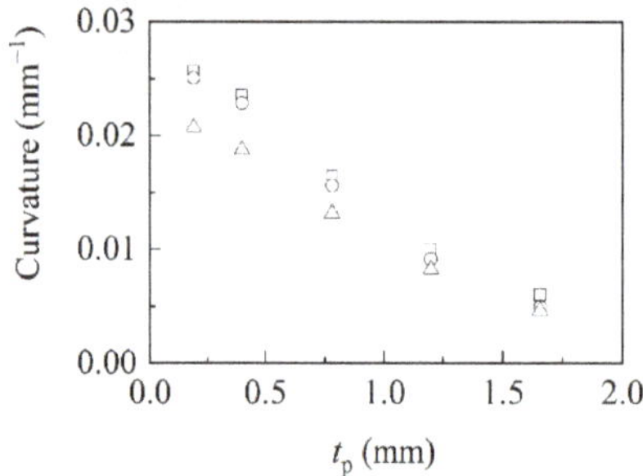

Figure A1. Computed curvatures for Laywood/PLA bilayers of experiment B in humid state using different modelling approaches. Open circles: Modeling and simulation based on layer-wise theory using ANSYS assuming linear elastic material. Open triangles: Modeling and simulation based on layer-wise theory using ANSYS assuming hyperelastic material. Open squares: Modeling based on classical laminate theory using the modification of Timoshenko's model in Equation (1). For highly curved samples with compressive strains, the more realistic hyperelastic model shows an increasing difference.

References

1. Oriani, A.; Scatena, V.L. The movement of involucral bracts of *Syngonanthus elegans* (Eriocaulaceae-Poales): Anatomical and ecological aspects. *Flora-Morphol. Distrib. Funct. Ecol. Plants* **2009**, *204*, 518–527. [CrossRef]
2. Evangelista, D.; Hotton, S.; Dumais, J. The mechanics of explosive dispersal and self-burial in the seeds of the filaree, *Erodium cicutarium* (Geraniaceae). *J. Exp. Biol.* **2011**, *214*, 521–529. [CrossRef]
3. Abraham, Y.; Elbaum, R. Hygroscopic movements in Geraniaceae: The structural variations that are responsible for coiling or bending. *New Phytol.* **2013**, *199*, 584–594. [CrossRef]
4. Elbaum, R.; Zaltzman, L.; Burgert, I.; Fratzl, P. The role of wheat awns in the seed dispersal unit. *Science* **2007**, *316*, 884–886. [CrossRef]

5. Dawson, C.; Vincent, J.F.V.; Rocca, A.-M. How pine cones open. *Nature* **1997**, *390*, 668. [CrossRef]
6. Poppinga, S.; Nestle, N.; Šandor, A.; Reible, B.; Masselter, T.; Bruchmann, B.; Speck, T. Hygroscopic motions of fossil conifer cones. *Sci. Rep.* **2017**, *7*, 40302. [CrossRef] [PubMed]
7. Elbaum, R.; Gorb, S.; Fratzl, P. Structures in the cell wall that enable hygroscopic movement of wheat awns. *J. Struct. Biol.* **2008**, *164*, 101–107. [CrossRef] [PubMed]
8. Fratzl, P.; Elbaum, R.; Burgert, I. Cellulose fibrils direct plant organ movements. *Faraday Discuss.* **2008**, *139*, 275–282. [CrossRef] [PubMed]
9. Fratzl, P.; Barth, F.G. Biomaterial systems for mechanosensing and actuation. *Nature* **2009**, *462*, 442–448. [CrossRef]
10. Zhang, S.; Zhou, S.; Liu, H.; Xing, M.; Ding, B.; Li, B. Pinecone-Inspired Nanoarchitectured Smart Microcages Enable Nano/Microparticle Drug Delivery. *Adv. Funct. Mater.* **2020**, *30*, 2002434. [CrossRef]
11. Reichert, S.; Menges, A.; Correa, D. Meteorosensitive architecture: Biomimetic building skins based on materially embedded and hygroscopically enabled responsiveness. *Comput. Aided Des.* **2015**, *60*, 50–69. [CrossRef]
12. Rüggeberg, M.; Burgert, I. Bio-inspired wooden actuators for large scale applications. *PLoS ONE* **2015**, *10*, e0120718. [CrossRef] [PubMed]
13. Vailati, C.; Hass, P.; Burgert, I.; Rüggeberg, M. Upscaling of wood bilayers: Design principles for controlling shape change and increasing moisture change rate. *Mater. Struct.* **2017**, *50*, 250. [CrossRef]
14. Erb, R.M.; Sander, J.S.; Grisch, R.; Studart, A.R. Self-shaping composites with programmable bioinspired microstructures. *Nat. Commun.* **2013**, *4*, 1712. [CrossRef] [PubMed]
15. Arslan, H.; Nojoomi, A.; Jeon, J.; Yum, K. 3D Printing of Anisotropic Hydrogels with Bioinspired Motion. *Adv. Sci.* **2019**, *6*, 1800703. [CrossRef] [PubMed]
16. Le Duigou, A.; Keryvin, V.; Beaugrand, J.; Pernes, M.; Scarpa, F.; Castro, M. Humidity responsive actuation of bioinspired hygromorph biocomposites (HBC) for adaptive structures. *Compos. Part A Appl. Sci. Manuf.* **2019**, *116*, 36–45. [CrossRef]
17. Le Duigou, A.; Castro, M. Moisture-induced self-shaping flax-reinforced polypropylene biocomposite actuator. *Ind. Crop. Prod.* **2015**, *71*, 1–6. [CrossRef]
18. Tahouni, Y.; Krüger, F.; Poppinga, S.; Wood, D.; Pfaff, M.; Rühe, J.; Speck, T.; Menges, A. Programming sequential motion steps in 4D-printed hygromorphs by architected mesostructure and differential hygro-responsiveness. *Bioinspir. Biomim.* **2021**. [CrossRef]
19. Correa, D.; Poppinga, S.; Mylo, M.D.; Westermeier, A.S.; Bruchmann, B.; Menges, A.; Speck, T. 4D pine scale: Biomimetic 4D printed autonomous scale and flap structures capable of multi-phase movement. *Philos. Trans. R. Soc. A* **2020**, *378*, 20190445. [CrossRef]
20. Tahouni, Y.; Cheng, T.; Wood, D.; Sachse, R.; Thierer, R.; Bischoff, M.; Menges, A. Self-shaping Curved Folding: A 4D-printing method for fabrication of self-folding curved crease structures. In *Symposium on Computational Fabrication*; Symposium on Computational Fabrication, Virtual Event USA, 05 11 2020 06 11 2020; Association for Computing Machinery: New York, NY, USA, 2020; Article 5; pp. 1–11. ISBN 9781450381703.
21. Reyssat, E.; Mahadevan, L. Hygromorphs: From pine cones to biomimetic bilayers. *J. R. Soc. Interface* **2009**, *6*, 951–957. [CrossRef]
22. Langhansl, M.; Dörrstein, J.; Hornberger, P.; Zollfrank, C. Fabrication of 3D-printed hygromorphs based on different cellulosic fillers. *Funct. Compos. Mater.* **2021**, *2*. [CrossRef]
23. Timoshenko, S. Analysis of bi-metal thermostats. *Josa* **1925**, *11*, 233–255. [CrossRef]
24. Wayne, R. *ImageJ*.; National Institues of Health: Stapleton, NY, USA, 2012.
25. Schindelin, J.; Arganda-Carreras, I.; Frise, E.; Kaynig, V.; Longair, M.; Pietzsch, T.; Preibisch, S.; Rueden, C.; Saalfeld, S.; Schmid, B.; et al. Fiji: An open-source platform for biological-image analysis. *Nat. Methods* **2012**, *9*, 676–682. [CrossRef] [PubMed]
26. Dave, M. Calculating Properties of a Triangle's Circumcircle. Available online: https://bitbucket.org/davemason/threepointcircumcircle/src/master/ (accessed on 12 July 2021).

Article

Deformation Behavior of Elastomer-Glass Fiber-Reinforced Plastics in Dependence of Pneumatic Actuation

Mona Mühlich [1,*], Edith A. González [1], Larissa Born [2], Axel Körner [1], Lena Schwill [2], Götz T. Gresser [2,3] and Jan Knippers [1]

[1] Institute of Building Structures and Structural Design (ITKE), University of Stuttgart, Keplerstrasse 11, 70174 Stuttgart, Germany; edith-anahi.gonzalez-san-martin@itke.uni-stuttgart.de (E.A.G.); axel.koerner@itke.uni-stuttgart.de (A.K.); jan.knippers@itke.uni-stuttgart.de (J.K.)

[2] Institute for Textile and Fiber Technologies (ITFT), University of Stuttgart, Pfaffenwaldring 9, 70569 Stuttgart, Germany; larissa.born@itft.uni-stuttgart.de (L.B.); lena.schwill@itft.uni-stuttgart.de (L.S.); goetz.gresser@itft.uni-stuttgart.de (G.T.G.)

[3] German Institutes of Textile and Fiber Research (DITF), Koerschtalstrasse 26, 73770 Denkendorf, Germany

* Correspondence: mona.muehlich@itke.uni-stuttgart.de

Abstract: This paper aims to define the influencing design criteria for compliant folding mechanisms with pneumatically actuated hinges consisting of fiber-reinforced plastic (FRP). Through simulation and physical testing, the influence of stiffness, hinge width as well as variation of the stiffness, in the flaps without changing the stiffness in the hinge zone, was evaluated. Within a finite element model software, a workflow was developed for simulations, in order to infer mathematical models for the prediction of mechanical properties and the deformation behavior as a function of the aforementioned parameters. In conclusion, the bending angle increases with decreasing material stiffness and with increasing hinge width, while it is not affected by the flap stiffness itself. The defined workflow builds a basis for the development of a predictive model for the deformation behavior of FRPs.

Keywords: fiber-reinforced plastic (FRP); elastomer; pneumatics; actuation; folding; finite element modeling (FEM); simulation

Citation: Mühlich, M.; González, E.A.; Born, L.; Körner, A.; Schwill, L.; Gresser, G.T.; Knippers, J. Deformation Behavior of Elastomer-Glass Fiber-Reinforced Plastics in Dependence of Pneumatic Actuation. *Biomimetics* **2021**, *6*, 43. https://doi.org/10.3390/biomimetics6030043

Academic Editors: Thomas Speck and Olga Speck

Received: 30 April 2021
Accepted: 17 June 2021
Published: 22 June 2021

Publisher's Note: MDPI stays neutral with regard to jurisdictional claims in published maps and institutional affiliations.

1. Introduction

According to the European Commission, buildings are accountable for around 40% of the energy consumption in the European Union. Strategies to improve the energy performance of the built environment include the integration of building technologies that aim to regulate indoor comfort [1].

Façades are the interface between the interior and exterior of a building. Therefore, by responding to external climates and regulating the internal conditions, they become an essential element in creating energy-neutral buildings [2]. Since external environmental conditions and internal occupancy patterns are in constant change, the façade's capability of adaptation plays a crucial role in improving the energy and environmental performance of a building.

Implementation of adaptive façades in Europe could lead to a 10% reduction of oil source energy and save 111 million tons of carbon dioxide per year [3]. On this account, the European Union's vision of nearly zero-energy buildings recognizes the potential of adaptive façades as a positive influence on a building's energy consumption [2,4].

Common implementations of adaptive façades rely on rigid body mechanisms to achieve the desired motion. However, in past years, less complex and promising solutions have derived from biomimetic principles. Prime examples are the current developments of adaptive façade elements that employ bio-inspired compliant mechanisms to drive kinematic behavior in the form of elastic deformation.

1.1. Previous Work

An early and well-known project based on this concept is flectofin®, inspired by the bird-of-paradise flower (*Strelitzia reginae*). Two lamellas consisting of fiber-reinforced plastic (FRP) are fixed to a stiff backbone and can bend up to 90° by applying force to the backbone. Novel at its time, this proved to be relatively limited in its kinematic behavior and fatigue strength due to the proportion of the modules to the actuators and the use of material with almost homogeneous stiffness that leads to high-stress concentrations in the FRP and high actuation forces [5].

Subsequent projects tried to overcome the limited fatigue strength by focusing on the development of folding elements with distinct flexible hinge zones. Flectofold was the first demonstrator to attempt this. For this project, the motion principle of *Aldrovanda vesiculosa* was abstracted into a curved-line folding mechanism for motion amplification with discrete curved hinge zones. The inflation of a pneumatic cushion located on the back of the midrib leads to bending of the midrib, and thereby to the amplified folding motion of the adjacent flaps which were connected to the midrib via hinge zones with reduced stiffness [6].

Current projects, such as the façade shading demonstrator Flexafold, have been inspired by the hindwings of insects. The wing-folding pattern of *Graphosoma lineatum italicum* was identified as flexagon principle and served as a basis for the geometry and folding pattern of Flexafold. Studies on the material set-up of hollow wing veins of the beetle *Dorcus titanus platymelus* were transferred into a FRP with a stiffness gradient from a flexible hinge zone to stiffer areas subjected to angular displacement, enabling the technical implementation of the Flexafold demonstrator [7,8].

The most recent example of the research work in this field is the ITECH Research Demonstrator 2018–19, which showcased the possibility of industrial manufacturing of large-scale folding elements. The motion is actuated by pneumatic cushions integrated into distinct flexible hinge zones. Taking as inspiration the folding behavior of ladybug wings (*Coleptera coccinellidae*), the zones next to the hinge areas increase in thickness, creating a reinforcement with higher stiffness in the critical zones [9].

1.2. Biomimetic Background

The research work presented is linked directly to the described research projects and demonstrators. Both the materiality and the actuation principle have been abstracted from the motion and actuation behavior of insect wings: in many species, the folding motion is realized by fibrous material which is actuated by internal pressure change.

The abstraction process, which underlies the present work, is shown in Figure 1. In particular, the ladybug wing (Figure 1a) was investigated as biological role model. The fragile hind wings of the beetle are shown in the scanning electron microscope (SEM) image (Figure 1b).

Two principles of the insect wing were abstracted and transferred into a demonstrator (Figure 1g). First principle is the material set-up itself based on the pliable zones of the wing: Next to the local, compliant folding zones, respectively the hinge zones of the wing, there are reinforced, stiffer zones. The thickening is formed on one side of the wing. This determines the folding direction of this pliable zone in the wing. Figure 1c shows a simplified representation. Instead of using different materials in the biological role model, just one material is used for the stiff and flexible zones. This material concept was abstracted and transferred into a FRP (Figure 1d). The FRP consists of elastomer and glass fiber-reinforced plastic (GFRP), covered and sealed with a thermoplastic polyurethane (TPU) film. These three materials form one hybrid FRP in which the stiffness can be adjusted by the cross-section-related content of GFRP.

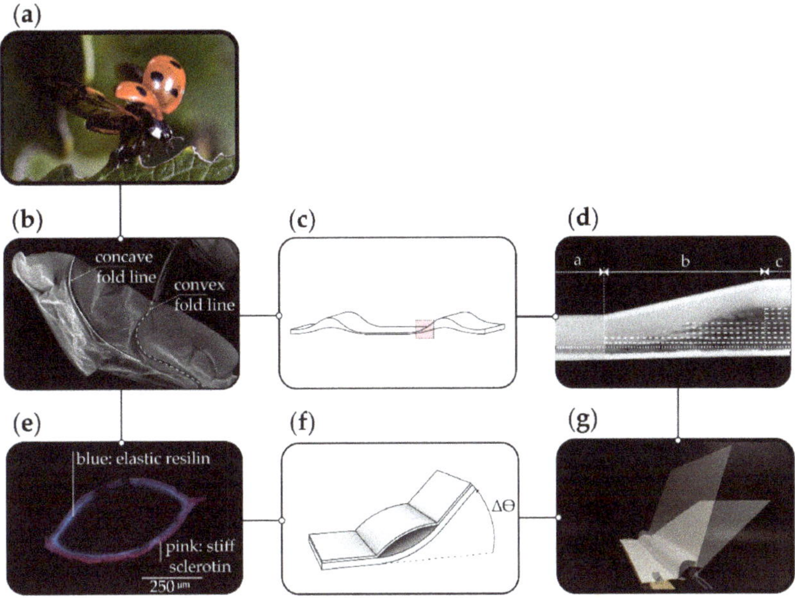

Figure 1. Biomimetic abstraction process with (**a**) biological role model *Coleoptera coccinellidae* [10], (**b**) scanning electron microscope image of the hindwing [11], (**c**) schematic of the material cross section in the insect wing (adapted from [7]), (**d**) hybrid FRP with graded material transitions between stiff and flexible zones [12], (**e**) autofluorescence image of the hollow vein in insect wing with marked sclerotin (pink) and resilin (blue) (adapted from [7]), (**f**) abstracted actuation principle with stiffer (bottom) and less stiff laminate and (**g**) FRP combining both principles (material and actuation).

The second principle is the folding mechanism based on the hollow wing veins (Figure 1e) [8]. The hollow wing is enclosed by resilin (flexible) and sclerotin (stiff) in different proportions on the upper and lower side of the hollow vein. This determines the bending direction when the vein is pressurized (Figures 1f and 2). Both principles were unified in a demonstrator shown in Figure 1g.

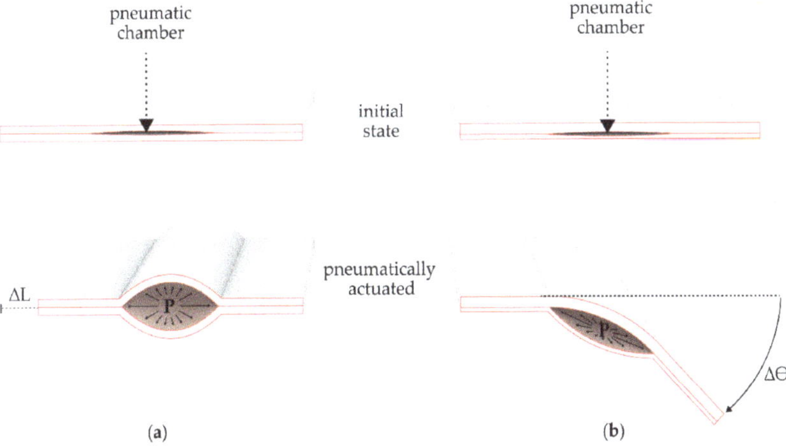

Figure 2. Actuation principle of elements with embedded pneumatic actuators with (**a**) plates with identical stiffness, leading to a shortening in length and (**b**) elements with plates of asymmetrical stiffness causing an angular displacement (adapted from [8]).

The demonstrator is the starting point for the investigations presented within the paper: the physical testing and simulation of FRP elements with asymmetric laminate set-ups. The embedded pneumatic actuation of the hinge will lead to an asymmetric elastic deformation, causing a rotation towards the less stiff side of the element (Figure 2). The aim of further exploring this kinematic principle is to create adaptive façade elements capable of achieving a high range of motions, yet robust enough to withstand the external forces acting on a building's skin.

2. Materials and Methods

2.1. Elastomer-Glass Fiber-Reinforced Plastic (E-GFRP) Hybrid Laminate

Based on preliminary research [8,13,14] regarding actuated FRPs, a combination of elastomer, GFRP and TPU film is used. The stiffness ratio between the laminate above and below the actuation chamber is adjusted by the GFRP content in the laminate set-up. Two intermediate elastomeric layers surrounding the cushion ensure the interlaminar adhesion between the upper and lower laminate once the actuation chamber expands. For protection against external environmental influences (UV radiation, water absorption, etc.), the laminate is coated with a TPU film. The material properties of the different layers are listed in Table 1.

Table 1. Material properties [12].

ID [1]	Material	Supplier	Trade Name	Fiber Orientation in °	Tensile Strength in MPa	Young's Modulus in MPa	Poisson's Ratio-
ET222	GFRP [2]	Toray Group Composite Materials (Italy) s.r.l.	ET222	0/90	546.50	16,628.61	0.20
HHZ99	Elastomer	Kraiburg GmbH & Co. KG	Kraibon HHZ9578/99	-	1.89	15.98	0.30
TPU	TPU [3]	Gerlinger Industries GmbH	4980	-	7.16	126.43	0.35

[1] short name within the paper, [2] glass fiber-reinforced plastic, [3] thermoplastic polyurethane.

Figure 3 shows an example of a laminate set-up of such an actuatable FRP. In the figure, the stiffness of the laminate above the actuation chamber is lower. The expansion of the chamber would cause the entire laminate to bend accordingly in the direction of this laminate.

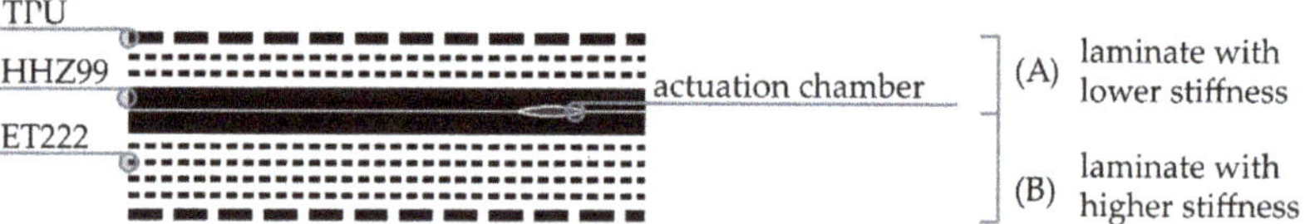

Figure 3. Schematic of an exemplary laminate set-up (TPU/ET222$_4$/HHZ99$_2$/ET222$_2$/TPU), where ET222 is implemented with a fiber orientation of ±45°.

2.2. Simulation of E-GFRP Hybrid Laminates

The simulations are based on the third-order theory with a stiffness update after each iteration. For the simulation workflow a three-dimensional model of the system is developed as input geometry for the nonlinear FEM simulation (SOFiSTiK Version 2018, SOFiSTiK AG, Oberschleißheim, Germany). The parametric model is discretized into quad shell elements and the geometry converted into elements with the needed information about, e.g., beams, couplings and supports for SOFiSTiK.

Also, the physical material characteristics set the boundaries for the simulations. First, the material behavior was calculated with the program Helius Composite 2017 by Autodesk. The composite type, the manufacturing properties and the fiber direction of each layer were set in order to calculate the characteristics of the compound materials and to define the thickness of the material in each area. Secondly, the interlaminar adhesion strength of the materials predetermined the maximal pressure for the actuation. The data basis for the materials was taken from Born's previous work [12].

To compare the simulation to physical tests, a consistent geometry of 50 cm × 30 cm was chosen for both physical and digital simulation. The geometry was then modeled in the FEM software SOFiSTiK. The element was fixed in all degrees of freedom on one side, to set it up as in the physical tests where it was clamped likewise. The fixation had an offset of 2 cm to the start of the hinge (Figure 4a,b).

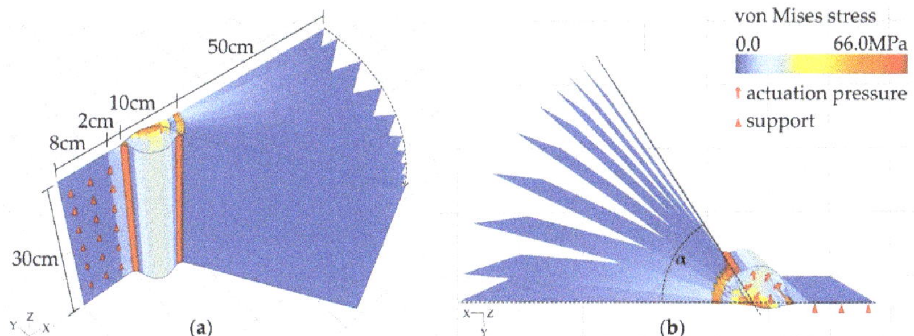

Figure 4. Simulation principle shown on test with 2 + 4 layers and a hinge width of 10 cm, (**a**) perspective view and (**b**) top view.

The tests investigated the influence of three different parameters on the resulting bending angle. The maximal actuation was defined by the maximal adhesion strength of the core material around the hinge, which for all tests was the elastomer HHZ99 with F_max = 477 N/10 cm [12].

2.3. Pneumatic Actuators

The motion is actuated by pneumatic cushions integrated within the hinge zone. The pneumatic actuators are fabricated with a 70 den (170 g/m^2) coated Nylon. This is a lightweight and airtight fabric that is coated on one side with a thermoplastic polyurethane that allows it to be heat sealable. The pneumatic actuators' fabrication starts by creating the desired shape with a PTFE film; this is later placed between two layers of TPU-coated nylon and heat pressed (Figure 5a).

This process will bond the edges surrounding the PTFE, resulting in a pneumatic cushion that can be inflated and deflated as needed. The shape of the cushion follows the dimension of the hinges created on the FRP specimens. However, in all instances the cushions are made 2 cm wider than the hinge, to ensure that they will fill in the volume created during actuation. The corners of the cushions are rounded, and a thin linear piece extends to create the cushion's inlet. For the air inlet, a pneumatic push-in connection is secured with an ear clamp to make it airtight. As previously mentioned, the composite elements are pressed with a piece of Teflon film in between layers. This creates a chamber in the component that allows for the insertion of the cushion. The cushion is inserted by hand, making sure it is evenly distributed through the chamber (Figure 5).

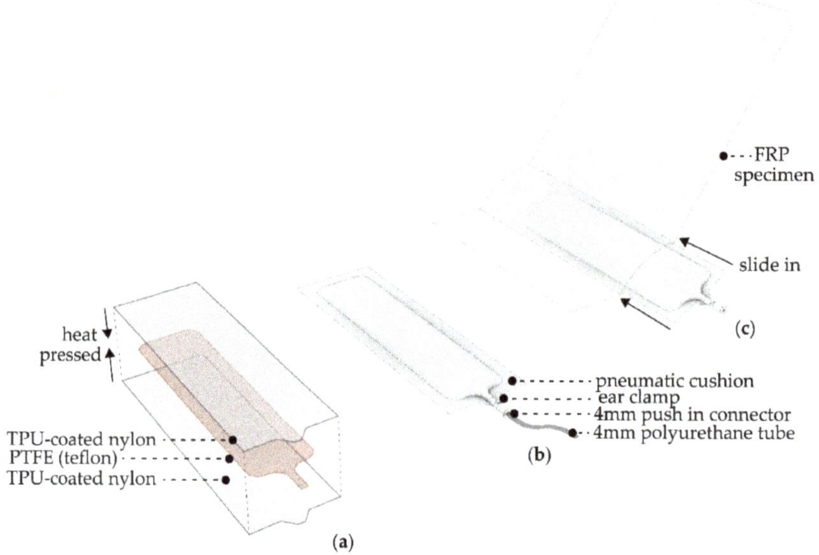

Figure 5. Configuration of pneumatic actuator exemplifying (**a**) materiality, (**b**) pneumatic fixtures and (**c**) integration of pneumatic cushion to composite specimen.

For this research, the cushions were pressed with two different techniques: with an industrial heat-press and manually with a regular flat iron. It was observed that the cushions made with the industrial heat-press proved to be less prone to leakage and more durable due to the absence of human error and the systematic production with defined pressure ranges, temperature, and time of pressing (Figure 6).

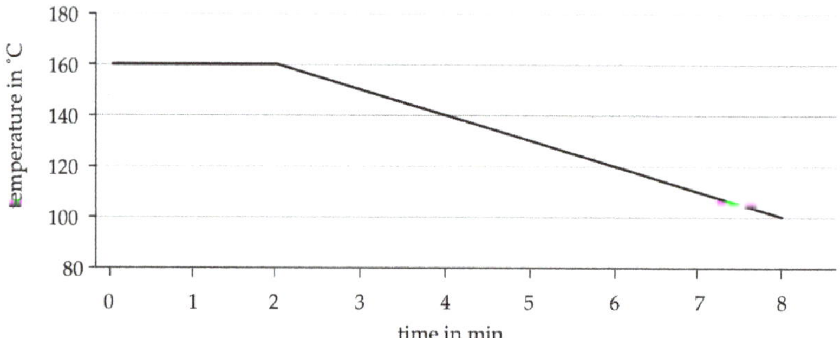

Figure 6. Temperature curve in pressing cycle at constant pressure of 1.3 bar.

2.4. Actuation and Bending Test Set-Up

The specimens were tested in a set-up that would allow the angular displacement of the element. The test set-up was designed to assess the specimens vertically to prevent gravitational forces from interfering in the results (Figure 7). The aim of these tests was to extract the air pressure needed in the pneumatic actuator for different bending angles and subsequently analyze the performance of the material.

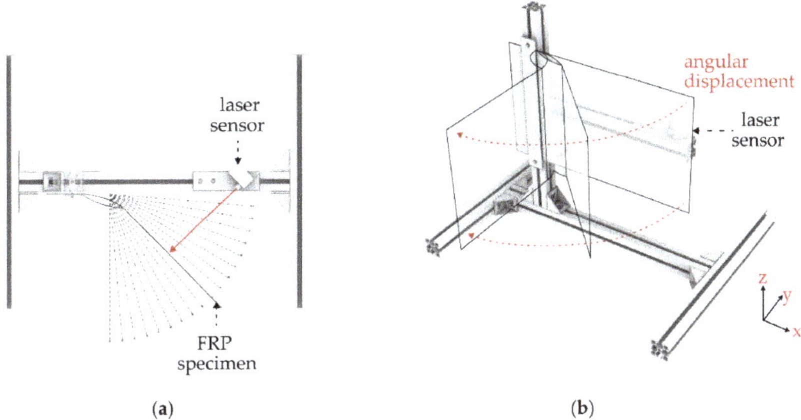

(**a**)   (**b**)

Figure 7. Vertical test set-up, with specimen clamped on one side to allow for the angular displacement (**a**) top view and (**b**) axonometric.

The test set-up consists of a clamping area in which the specimen is fixed at one edge. When the integrated pneumatic cushion is actuated, the asymmetric material layup will enable the bending of the hinge, causing an angular displacement towards the direction of the thinner layer. A distance laser sensor placed on the opposite side of the clamping area will give the distance to the element. With this measurement, the element's final bending angle can be calculated by using a set of trigonometric functions. The measurement errors have not been considered, because they are mostly systematic and for the comparison of the results within our set up thereby of less impact (Figure 8) [15].

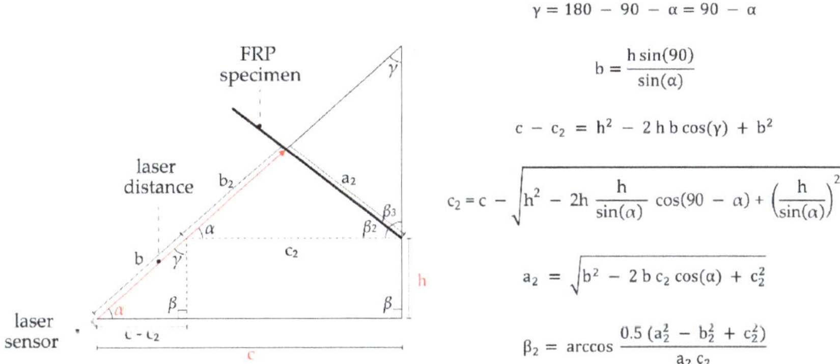

$$\gamma = 180 - 90 - \alpha = 90 - \alpha$$

$$b = \frac{h\sin(90)}{\sin(\alpha)}$$

$$c - c_2 = h^2 - 2\,h\,b\cos(\gamma) + b^2$$

$$c_2 = c - \sqrt{h^2 - 2h\,\frac{h}{\sin(\alpha)}\cos(90 - \alpha) + \left(\frac{h}{\sin(\alpha)}\right)^2}$$

$$a_2 = \sqrt{b^2 - 2\,b\,c_2\cos(\alpha) + c_2^2}$$

$$\beta_2 = \arccos\frac{0.5\,(a_2^2 - b_2^2 + c_2^2)}{a_2\,c_2}$$

Figure 8. Trigonometric functions to find the angular displacement, when h, c and α are the input variables (adapted from [15]).

The system works by setting up the desired bending angle; this enables a compressor that inflates the pneumatic cushion until the desired angle is achieved. After running the tests for the required number of cycles, one can extract the reading of the bending angles and the exact air pressure needed to reach them.

3. Results

3.1. Simulative Investigation of the Geometrical Parameters of an Actuated Hinge

For the following results, we changed the stiffness, the hinge width and the material in the surface area, compared to the hinge, taking minimal three variations into consideration. An overview of the tests and their compositions can be seen in Table 2.

Table 2. Material test overview.

ID [1]	Hinge Width in cm	Continuous Layers ET222 (±45)°		Width HHZ99 in cm	Layers ET222 (0/90)° (Replacing HHZ99) [3]
		Laminate (A) [2]	Laminate (B) [2]		
PPE_2 + 4	10	2	4	continuous	-
PPE_4 + 8	10	4	8	continuous	-
PPE_6 + 12	10	6	12	continuous	-
PPE_2 + 4_G5	5	2	4	continuous	-
PPE_2 + 4_G7.5	7.5	2	4	continuous	-
PPE_2 + 4_G10	10	2	4	continuous	-
PPE_2 + 4_G12.5	12.5	2	4	continuous	-
PPE_2 + 4_G15	15	2	4	continuous	-
PPE_2 + 4_E15 + 5	10	2	4	15 (+5)	4 layers ET222
PPE_2 + 4_E17.5 + 5	10	2	4	17.5 (+5)	4 layers ET222
PPE_2 + 4_E20 + 5	10	2	4	20 (+5)	4 layers ET222

[1] short name within the paper; [2] cf. Figure 3 (lower stiffness in laminate (A) and higher stiffness in laminate (B)); [3] cf. Figure 11 (stiffening of the flaps by substituting HHZ99 with four layers of ET222 with a fiber orientation of (0/90)° in relation to the bending axis).

The first simulations compare the bending angles of three different total stiffnesses: two layers of ET222 on one side of the hinge and four on the other side and, respectively, four and eight layers as well as six and twelve layers. The thinnest lamina resulted in the largest bending angle of the element, and therefore was chosen for the subsequent tests. The three maximal bending angles suggest a negative exponential correlation between material thickness and bending angle (Figure 9).

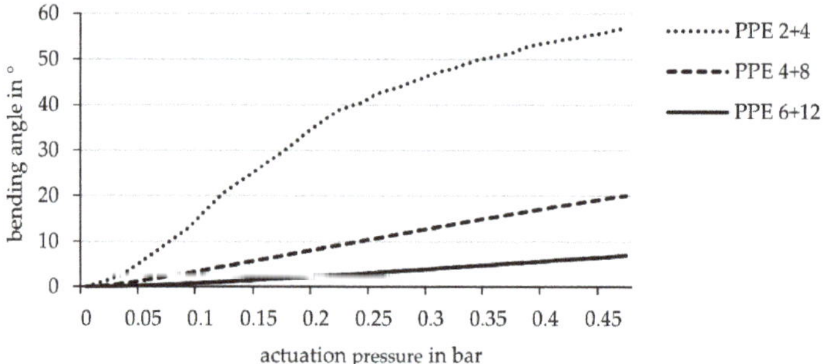

Figure 9. Achievable bending angle (simulation) in relation to the actuation pressure for laminate set-ups with different stiffness's: 2 + 4 layers ET222 (laminate set-up of PPE 2 + 4 similar to Figure 2), 4 + 8 layers ET222 and 6 + 12 layers ET222, hinge width 10 cm.

The second series of tests investigate the influence of hinge width. The investigated hinge widths are based on experience values from previous research work and experiments [8,9]. The hinge width was increased by 2.5 cm steps, ranging from 5 cm to 15 cm width. The results showed a big divergence of in the maximal bending angle between the 5 cm (15.3°), the 7.5 cm (40.7°) and the 10 cm wide hinge (57°), while the difference from the 12.5 cm width (66.4°) to the 15 cm wide hinge (70.2°) was comparatively small (Figure 10).

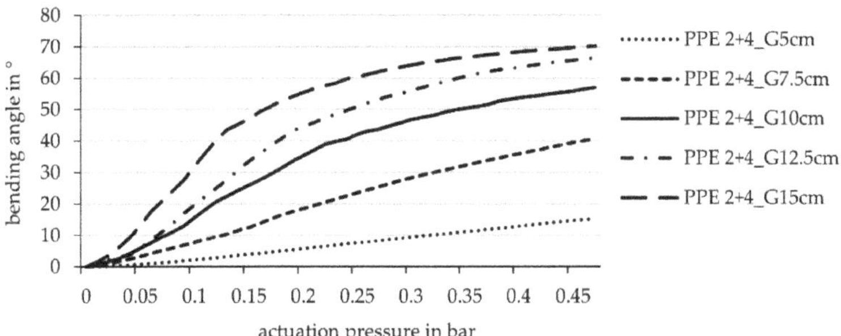

Figure 10. Achievable bending angle (simulation) in relation to the actuation pressure for laminate set-ups with different hinge widths: 5 cm, 7.5 cm, 10 cm, 12.5 cm, 15 cm (laminate set-up PPE 2 + 4 cf. Figure 1 equal for all samples).

In the third test, the elastomer in the flaps is replaced by four layers of ET222, which corresponds to the thickness of one layer of HHZ99. The orientation of these ET222 layers is shifted by 45° (compared to the continuous ET222 layers), resulting in an effective fiber direction of $(0/90)°$ in relation to the bending axis (Figure 11).

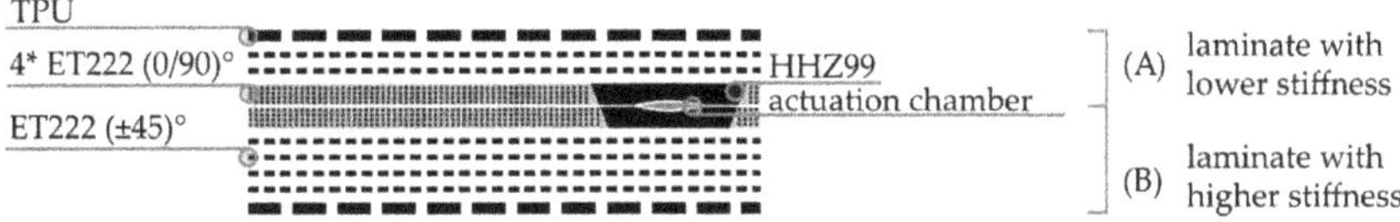

Figure 11. Schematic sketch of laminate set-up with stiffened flaps due to the substitution of HHZ99 by four layers ET222 (fiber orientation 0/90°) (HHZ99 layer on laminate side (**A**) with lower stiffness 5 cm wider than on laminate side (**B**) with higher stiffness).

In the simulation, variations of the elastomer width around the hinge were calculated with the set-up of the thin lamina and 10 cm hinge width. The difference of the bending angle proved to be very small. However, compared to the previous results for the thin lamina with a 10 cm hinge width, the actuation curve starts steeper and ends at a slightly higher bending angle (Figure 12).

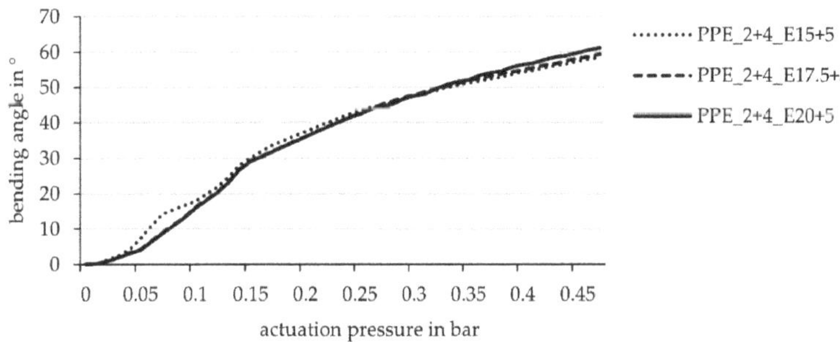

Figure 12. Achievable bending angle (simulation) in relation to the actuation pressure for laminate set-ups with different widths of the elastomeric layers enclosing the cushion chamber (width 10 cm): 15 + 5 cm, 17.5 + 5 cm, 20 + 5 cm (elastomer layer on the side with lower stiffness 5 cm wider than on the side with higher stiffness, cf. Figure 11).

3.2. Mechanical Investigation of the Geometrical Parameters of an Actuated Hinge

In addition to the simulative, theoretical examination, the abovementioned laminate set-ups were also tested mechanically by means of test components. The size of the flap of the specimens to be actuated was 30 × 30 cm. It can be noted that the maximum deformation angle of the specimens almost matches the simulation results (Figure 13).

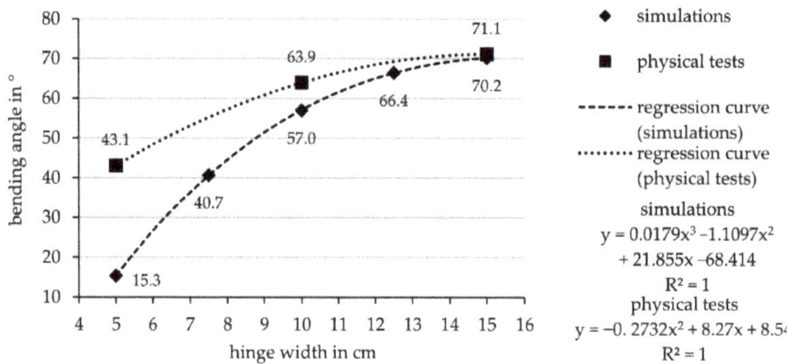

Figure 13. Regression formula with a polynomic approximation curve based on the results of the simulations and the physical tests.

However, the deformation angle resulting from the simulation is generally smaller than that of the real mechanical test, because the physical tests have been actuated till the failure of the material, which emerged to be a higher pressure than in the simulations. This is caused by the data regarding the interlaminar adhesion on which the simulation and the failure criteria are based. The data on interlaminar adhesion result from T-peel tests in which the plies are peeled off from each other at a 90° angle. However, when the vacuum chamber is actuated, a much smaller angle is created between the plies, visible in the larger discrepancies of the smaller hinges where higher pressures than expected, could be applied. Thus, the limit chosen for the simulations is relatively low and probably can be set higher. Thereby, also the actuation pressure and bending angle would be higher in the simulations and the results of simulation and practical testing would converge.

3.3. Component Actuated with Multiple Cushions

Based on the investigations presented in Section 3.1, it can be stated that a bending angle of 90° cannot be achieved with only one cushion. A larger bending angle could generally be achievable by combining several cushions next to each other. The regression analysis of the data of the previous test specimens (variation of the hinge width) should predict the necessary hinge width for a deformation of 90°.

3.3.1. Regression Model from Simulative Investigation

For the regression formula (Figure 13), a polynomic approximation curve was used as it fitted the results best; with the formula predictions about the resulting bending angles for varies hinge widths were made (Table 3).

The interpolation for the angle of 45° yielded a hinge width of 8.14 cm. Hence, for the simulation of the deformation behavior, two hinges with a hinge width of 8 cm were used to reach a bending angle of approximately 90°. The results of the simulation regarding the deformation behavior of a sample with two cushions and a hinge width of 8 cm are shown in Figure 14.

Table 3. Bending angle dependent of hinge widths (simulation) and prediction of bending angle/hinge width (regression formula from Figure 13 used).

Simulations		Prediction Based on Diagram			
Hinge Width in cm	Angle in °	Hinge Width in cm	Angle in °	Angle in °	Hinge Width in cm
5	15.3	6	26.3	35	6.92
7.5	40.7	7	35.7	40	7.50
10	57.0	8	**44.0**	**45**	**8.14**
12.5	66.4	9	51.1	50	8.84
15	70.2	10	57.1	55	9.63

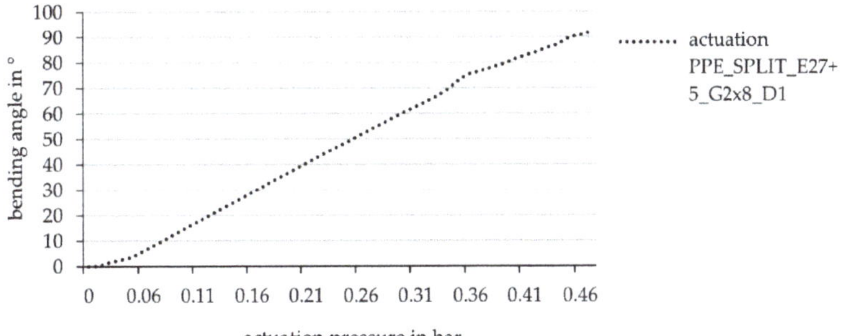

Figure 14. Results of simulative testing of an element with two hinges (hinge width 8 cm, distance between hinges 1 cm), bending angle in relation to actuation pressure.

It can be said that the bending angle of 90° is achieved, and thereby also that the simulative data basis is sufficient to predict the deformation behavior, respectively, the bending angle of a FRP with the described and investigated material set-up.

3.3.2. Validation of Simulative Investigation via Physical Tests

Because of the aforementioned conservative boundary conditions set for the actuation and failure mechanism in the simulations, the resulting maximal bending angles of the physical tests are higher with less hinge width (Table 4). Calculating similar to the simulation would result in two hinges with a width of 5 cm. Considering the hinge width of 8 cm of the simulation results, the mean value is: 6.5 cm hinge width.

Table 4. Bending angle dependent of hinge widths (physical tests) and prediction of bending angle/hinge width (regression formula from Figure 13 used).

Physical Tests		Prediction Based on Diagram			
Hinge Width in cm	Angle in °	Hinge Width in cm	Angle in °	Angle in °	Hinge Width in cm
5	43.1	4	37.2	35	3.64
10	63.9	4.5	40.2	40	4.46
15	71.1	**5**	**43.1**	**45**	**5.36**
		6	58.3	50	6.34
		6.5	**50.7**	**55**	**7.45**

Finally, for the validation of the results derived from the simulation, a test specimen with two side-by-side hinges, each with a width of 6.5 cm and a distance between them of 1 cm, was manufactured and tested in the test rig (cf. Figure 15).

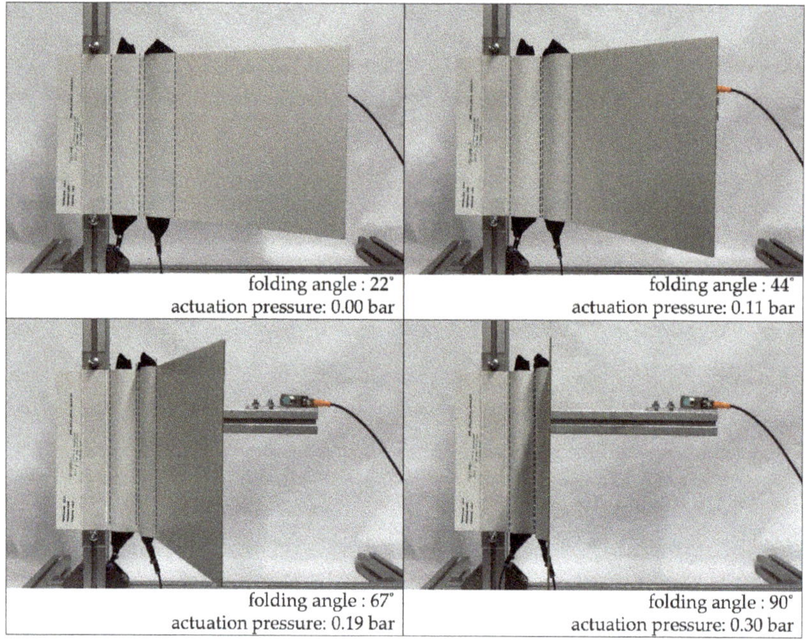

Figure 15. Mechanical testing of the hinge component with two hinges (hinge width 6.5 cm, distance between the hinges 1 cm).

The maximum deformation angle is 90.68° ± 0.47° at a pneumatic pressure of 0.33 bar ± 0.01 bar (Figure 16). The statistical calculation of mean value and standard deviation is based on the first 38 cycles of testing after 150 load cycles, the component delaminated in the bridging area between the two cushion chambers; however, not between the two HHZ99 layers, but between HHZ99 and ET222 (stiff component side).

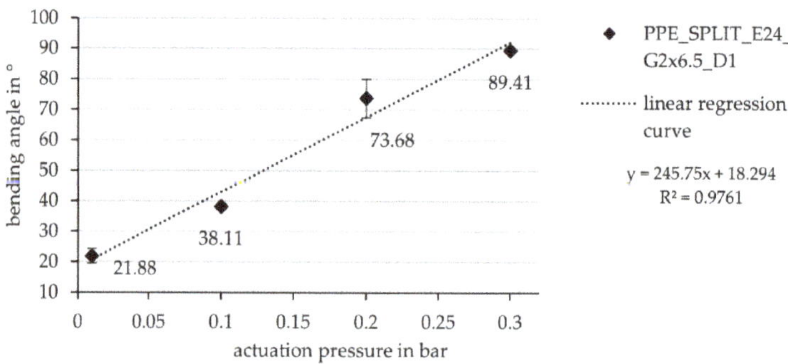

Figure 16. Bending angle as a function of the actuation pressure for the hinge component with two cushions (width per hinge 6.5 cm; distance between hinges 1 cm).

Thus, the simulation can be confirmed by the tests. It is possible to adjust the deformation angle via the geometric framework conditions of the component. However, the fatigue strength must still be improved for the technical application.

4. Discussion

Both the simulation and the physical experiments demonstrate the adjustability of the mechanical properties by adapting the geometry on the basis of the derived regression functions. However, there is a need for further development with regard to fatigue strength, since material failure due to interlaminar adhesion already occurs after 150 cycles. The interlaminar adhesion of this material structure could be improved, for example, by sewing the individual layers together or by Z-pinning. Z-pinning hereby means to introduce small metal or FRP pins through the thickness of the textile stack [16,17]. In contrast to sewing, this can also be done with pre-impregnated fabrics like ET222. A non-material-homogeneous connection between the layers has already been successfully applied in the ITECH Research Demonstrator 2018-19 in the form of screw connections next to the hinges [9]. In contrast to sewing, this can also be done with pre-impregnated fabrics like ET222. However, it would be better to adjust the interlaminar strength via the material properties so that sufficient interlaminar adhesion is present. Conceivably, such strength is achievable by using multilayered woven fabrics, since here the layers are already connected in the Z-direction. 3D woven structures would accordingly offer a possibility to produce components with higher fatigue strength without additional process steps. Additionally, another possibility to avoid the problem of delamination could be an actuation that is distributed over a larger surface compared to the current rod-shaped one. The better pressure distribution due to the larger surface could reduce the load on individual parts. Here, the dynamics of interlaminar adhesion and the geometry of the actuation would have to be further investigated.

The empirical simulations and studies can be used as a data basis for designing new adaptive components and as a benchmark for further investigations. The functionality is generally proven, but the bending angle should be increased for use in various application scenarios. The authors see this research on the bending behavior of a simple geometry as an important step towards the development of adaptive bending-active façade systems. These systems have the potential to be a feasible alternative to conventional mechanical shading systems for complex building geometries. This research contributes to the development of an ongoing project of an adaptive façade system to be installed at the University of Freiburg. This adaptive, bioinspired shading device will consist of FRP elements with asymmetric laminate set-ups and variable stiffness, thereby creating linear hinge zones which will be activated by integrated pneumatic actuators. The façade is being developed by the University of Stuttgart and is intended to be completed by 2022.

Author Contributions: Conceptualization, M.M., E.A.G. and L.B.; methodology, M.M., E.A.G. and L.B.; software, M.M. and E.A.G.; validation, M.M., E.A.G. and L.B.; investigation, M.M., E.A.G. and L.B.; writing—original draft preparation, M.M., E.A.G., L.B., A.K. and L.S.; writing—review and editing, G.T.G. and J.K.; visualization, M.M., E.A.G. and L.B.; supervision, G.T.G. and J.K.; project administration, G.T.G. and J.K.; funding acquisition, G.T.G. and J.K. All authors have read and agreed to the published version of the manuscript.

Funding: This research was funded by the Ministry of Science, Research and Arts Baden-Württemberg (MWK) within the sub-project BioElast_01 "TRR 141-A03: Force transmission and actuation in the transition zone between rod-shaped and flat component elements" of the cooperation project "Bio-inspired elastic material systems and composite components for sustainable building in the 21st century".

Acknowledgments: The authors would like to thank the biologists at the Universities of Tübingen and Freiburg for the constructive cooperation work and the insight into the fascinating and complex world of biology, which leads often to a deeper understanding of technical mechanics and to creative new developments. The authors would also like to acknowledge Jenin Gerlach and Frederik Wulle form the Institute for Control Engineering of Machine Tools and Manufacturing Units (ISW) of the University of Stuttgart, for providing the software utilized to test the angular displacement of the modules and their guidance throughout the testing process. The work presented in this paper was partially supported by the Deutsche Forschungsgemeinschaft (DFG, German Research Foundation) under Germany's Excellence Strategy—EXC 2120/1—390831618.

Conflicts of Interest: The authors declare no conflict of interest.

References

1. Energy Performance of Buildings Directive: Energy Performance of Buildings Standards. Available online: https://ec.europa.eu/energy/topics/energy-efficiency/energy-efficient-buildings/energy-performance-buildings-directive_en#energy-performance-of-buildings-standards (accessed on 26 April 2021).
2. Loonen, R.C.; Favoino, F.; Hensen, J.L.; Overend, M. Review of current status, requirements and opportunities for building performance simulation of adaptive facades. *J. Build. Perform. Simul.* **2017**, *10*, 205–223. [CrossRef]
3. Barozzi, M.; Lienhard, J.; Zanelli, A.; Monticelli, C. The Sustainability of Adaptive Envelopes: Developments of Kinetic Architecture. *Procedia Eng.* **2016**, *155*, 275–284. [CrossRef]
4. Sandak, A.; Sandak, J.; Brzezicki, M.; Kutnar, A. *Bio-Based Building Skin*; Springer: Singapore, 2019; ISBN 978-981-13-3746-8.
5. Lienhard, J.; Schleicher, S.; Poppinga, S.; Masselter, T.; Milwich, M.; Speck, T.; Knippers, J. Flectofin: A hingeless flapping mechanism inspired by nature. *Bioinspir. Biomim.* **2011**, *6*, 45001. [CrossRef] [PubMed]
6. Körner, A.; Born, L.; Mader, A.; Sachse, R.; Saffarian, S.; Westermeier, A.S.; Poppinga, S.; Bischoff, M.; Gresser, G.T.; Milwich, M.; et al. Flectofold—a biomimetic compliant shading device for complex free form facades. *Smart Mater. Struct.* **2018**, *27*, 17001. [CrossRef]
7. Schieber, G.; Born, L.; Bergmann, P.; Körner, A.; Mader, A.; Saffarian, S.; Betz, O.; Milwich, M.; Gresser, G.T.; Knippers, J. Hindwings of insects as concept generator for hingeless foldable shading systems. *Bioinspir. Biomim.* **2017**, *13*, 16012. [CrossRef] [PubMed]
8. Mader, A.; Born, L.; Körner, A.; Schieber, G.; Masset, P.-A.; Milwich, M.; Gresser, G.T.; Knippers, J. Bio-inspired integrated pneumatic actuation for compliant fiber-reinforced plastics. *Compos. Struct.* **2020**, *233*, 111558. [CrossRef]
9. Körner, A.; Born, L.; Bucklin, O.; Suzuki, S.; Vasey, L.; Gresser, G.T.; Menges, A.; Knippers, J. Integrative design and fabrication methodology for bio-inspired folding mechanisms for architectural applications. *Comput. Aided Des.* **2021**, *133*, 102988. [CrossRef]
10. Coleoptera Coccinellidae. Available online: https://www.pexels.com/de-de/foto/roter-und-schwarzer-kafer-36485/ (accessed on 30 April 2021).
11. Hindwing Coleoptera Coccinellidae. Available online: https://www.itke.uni-stuttgart.de/de/forschung/icd-itke-forschungspavillons/itech-forschungsdemonstrator-2018-19 (accessed on 30 April 2021).
12. Born, L. Grundlagen für die Auslegung und Gestaltung eines Hybridmaterials für Außen Liegende, Adaptive Fassadenbauteile aus Faserverbundkunststoff. Ph.D. Thesis, University of Stuttgart, Stuttgart, Germany, 23 September 2020.
13. Born, L.; Körner, A.; Mader, A.; Schieber, G.; Milwich, M.; Knippers, J.; Gresser, G.T. Adaptive FRP Structures For Exterior Applications. *Adv. Mater. Lett.* **2019**, *10*, 913–918. [CrossRef]
14. Born, L.; Körner, A.; Schieber, G.; Westermeier, A.S.; Poppinga, S.; Sachse, R.; Bergmann, P.; Betz, O.; Bischoff, M.; Speck, T.; et al. Fiber-Reinforced Plastics with Locally Adapted Stiffness for Bio-Inspired Hingeless, Deployable Architectural Systems. *Key Eng. Mater.* **2017**, *742*, 689–696. [CrossRef]
15. Gerlach, J. Systementwicklung einer Steuerung und Regelung für Pneumatisch Aktuierte Verschattungselemente. Master's Thesis, University of Stuttgart, Stuttgart, Germany, 27 October 2020.
16. Mouritz, A.P. Review of z-pinned composite laminates. *Compos. Part A Appl. Sci. Manuf.* **2007**, *38*, 2383–2397. [CrossRef]
17. Pingkarawat, K.; Mouritz, A.P. Improving the mode I delamination fatigue resistance of composites using z-pins. *Compos. Sci. Technol.* **2014**, *92*, 70–76. [CrossRef]

Article

Self-Actuated Paper and Wood Models: Low-Cost Handcrafted Biomimetic Compliant Systems for Research and Teaching

Simon Poppinga [1,2,3,*,†], Pablo Schenck [1], Olga Speck [1,3], Thomas Speck [1,2,3], Bernd Bruchmann [4] and Tom Masselter [1,*,†]

1 Plant Biomechanics Group @ Botanic Garden, University of Freiburg, 79104 Freiburg im Breisgau, Germany; pabloschenck7@yahoo.de (P.S.); olga.speck@biologie.uni-freiburg.de (O.S.); thomas.speck@biologie.uni-freiburg.de (T.S.)
2 Freiburg Materials Research Center (FMF), University of Freiburg, 79104 Freiburg im Breisgau, Germany
3 Cluster of Excellence livMatS @ FIT—Freiburg Center for Interactive Materials and Bioinspired Technologies, University of Freiburg, 79110 Freiburg im Breisgau, Germany
4 BASF SE, Advanced Materials and Systems Research, 67056 Ludwigshafen/Rhein, Germany; bernd.bruchmann@basf.com
* Correspondence: simon.poppinga@biologie.uni-freiburg.de (S.P.); tom.masselter@biologie.uni-freiburg.de (T.M.)
† These authors contributed equally to this work.

Citation: Poppinga, S.; Schenck, P.; Speck, O.; Speck, T.; Bruchmann, B.; Masselter, T. Self-Actuated Paper and Wood Models: Low-Cost Handcrafted Biomimetic Compliant Systems for Research and Teaching. *Biomimetics* **2021**, *6*, 42. https://doi.org/10.3390/biomimetics6030042

Academic Editor: Jose Luis Chiara

Received: 11 May 2021
Accepted: 18 June 2021
Published: 22 June 2021

Publisher's Note: MDPI stays neutral with regard to jurisdictional claims in published maps and institutional affiliations.

Abstract: The abstraction and implementation of plant movement principles into biomimetic compliant systems are of increasing interest for technical applications, e.g., in architecture, medicine, and soft robotics. Within the respective research and development approaches, advanced methods such as 4D printing or 3D-braiding pultrusion are typically used to generate proof-of-concept demonstrators at the laboratory or demonstrator scale. However, such techniques are generally time-consuming, complicated, and cost-intensive, which often impede the rapid realization of a sufficient number of demonstrators for testing or teaching. Therefore, we have produced comparable simple handcrafted compliant systems based on paper, wood, plastic foil, and/or glue as construction materials. A variety of complex plant movement principles have been transferred into these low-cost physical demonstrators, which are self-actuated by shrinking processes induced by the anisotropic hygroscopic properties of wood or paper. The developed systems have a high potential for fast, precise, and low-cost abstraction and transfer processes in biomimetic approaches and for the "hands-on understanding" of plant movements in applied university and school courses.

Keywords: actuators; biomimetics; compliant systems; hygroscopic materials; plant movements

1. Introduction

1.1. Motivation

Although they do not possess muscles, nerves, and "typical" (technical rotating) hinges, plants can perform a variety of complex motions. The speed of movement spans several orders of magnitude and depends on the dimension of the motile plant structure, its mechanical properties, and the actuation principle [1]. Whereas slow motion is typically driven by hydraulic processes, i.e., water displacement processes between cells and tissues, fast motion is characterized by (additional) elastic instabilities, i.e., the release of elastic energy accumulated and stored in a deformed (prestressed) structure. Hydraulics include turgor changes, which require metabolic energy during the motion, and energetically cost-free hygroscopic swelling/shrinking processes (reviewed by [2–4]). Famous examples of plant movements are the touch-sensitive leaflets and leaves of *Mimosa pudica*, which move by turgor changes [5], the hygroscopic opening and closing of pine cones (*Pinus* spp.) [6], and the snapping of the carnivorous Venus flytrap (*Dionaea muscipula*), which incorporates both turgor changes and the release of prestress [7,8].

Motile plant structures change shape during movement and, therefore, constitute compliant systems [9,10]. This is in contrast to rigid-body mechanisms with "classical" hinges, as characterized by stiff structural components that glide against each other and that are typically connected by rotating hinges. Motile plant structures are of increasing interest for biomimetic approaches in which hinge-free and functionally resilient and robust compliant mechanisms are desired, e.g., for applications in architecture, soft robotics, and medicine [11–14]. Previously existing biomimetic motile technical structures incorporate nastic responses, which involve movement in a morphologically predetermined (structurally preprogrammed) manner after being triggered [15], or tropistic reactions in which the movement direction depends on the direction of the triggering stimulus [16]. The seed scale of the pinecone is a widely employed biological concept generator for nastic biomimetic motion. Its functional bilayer architecture, which consists of tissue layers with different passive swelling and shrinking properties [6,17,18], has been abstracted and transferred into numerous motile technical materials systems. These biomimetic structures can react with bending to various stimuli, e.g., changes in humidity, temperature, pH, or light, depending on the materials used for bilayer construction [19–21]. By tailoring the materials and their compositions and the architecture of the material layers, more complex motions can be achieved, such as flat-to-helical transitions and coiling [22,23]. Three-dimensional printing of shape-changing structures (termed 4D printing) has become an emerging field [24–26], allowing the transfer of complex plant deformation principles such as multi-phase motion and curved-fold bending into 4D prints [27,28].

However, 3/4D printing and other manufacturing methods are costly, complicated, and time-consuming, all of which might hinder the rapid realization of a sufficient number of demonstrators for testing or for teaching students. Therefore, we have conducted a search for a method for the construction of low-cost, biomimetic, self-actuated compliant systems out of common, cost-effective materials. These systems should be capable of performing physically correct nastic plant deformation and motion sequences. In prior studies, we have presented handcrafted structures that mimic the movement of the bird of paradise flower (*Strelitzia reginae*) [29] and the trap movements of various carnivorous plants [30]. However, these systems are actuated extrinsically by the application of mechanical force (e.g., by manual bending) and are, therefore, not self-actuated. For the new self-actuated structures presented in this study, we have aimed at implementing passive hydraulic actuation as the main driver. Therefore, we have chosen paper and wood as basic construction materials, as these are characterized by pronounced swelling/shrinking capabilities upon hydration/dehydration [31,32]. The use of connected wood and/or paper strips with different fiber orientations to generate hygroscopically driven bending and curling motions has a long history and has also found applications in didactic approaches [33]. However, spatially complex movements have not been achieved to date. Our goal has been to transfer the deformation principles of various plant concept generators into simple and low-cost demonstrators.

1.2. Plant Concept Generators

1.2.1. Flower of the Bird of Paradise (*Strelitzia reginae*)

The flower of the bird of paradise (*S. reginae*) (Figure 1a) is characterized by a motile perch-like sheath that plays a crucial role in pollination [34,35]. When a bird lands on the sheath, which consists of two connected petals forming a cylinder open on one longitudinal side, to search for nectar at the flower base, its body weight causes a downward bending and the simultaneous sideways flapping of the two petals by torsional buckling [36]. Thereby, the previously hidden stamens are exposed, and the pollen can stick to the bird's feet for transport to and pollination of another flower. When the bird flies away, the sheath closes, and the stamens with pollen are protected by the sheath once again. The deformation principle (torsional buckling) has been abstracted and implemented in the biomimetic façade shading Flectofin® [36]. The sheath movement is reversible and highly repeatable, and is not self-actuated. In our approach, which we present here, we have aimed at producing a self-actuated technical "counterpart" of the deformation behavior.

Figure 1. Biological concept generators for the compliant systems presented in this study. (**a**) The flower of the bird of paradise (*S. reginae*) features a motile sheath that opens upon application of a mechanical load. (**b**) The false rose of Jericho (*S. lepidophylla*) is in a strongly folded configuration when dry and unfolds during wetting. (**c**) The snap-trap of the carnivorous aquatic waterwheel plant (*A. vesiculosa*) closes after mechanical triggering of sensitive hairs inside the trap (not visible). The two trap lobes are connected by a midrib (arrow) and close because of the slight bending of the midrib but without curvature changes of the two trap lobes. Image modified after [37]. (**d**) The lily (*L.* spec.) flower opens via edge-growth-based actuation of the individual petals. (**e**) The open seed pod of the Spanish broom (*S. junceum*) is an example of the hygroscopic seed pod of members of the Fabaceae. (**f**) The snap-trap of the carnivorous Venus flytrap (*D. muscipula*) closes after the mechanical triggering of sensitive hairs (black arrow), leading to a swift concave-convex curvature change of the two trap lobes.

1.2.2. False Rose of Jericho (*Selaginella lepidophylla*)

The false rose of Jericho (*S. lepidophylla*) is a spike-moss capable of withstanding severe droughts and pronounced water loss (Figure 1b). Under dry environmental conditions, each of its stems is curved inward, and the whole plant has a spherical appearance [38].

Under wet conditions, the stems and leaves soak up water and unfold. The reversible hygroscopic folding/unfolding takes place in a coordinated manner, with the individual stems reacting as individual elements and is dictated by the bilayer architecture of the stems [39]. Here, our aim has been to develop a paper-based composite structure consisting of individual hygroscopic elements (the "stems") that act as a functional unit and induce the global opening and closing response of the whole artificial "plant".

1.2.3. Snap-Trap of the Carnivorous Waterwheel Plant (*Aldrovanda vesiculosa*)

The aquatic carnivorous waterwheel plant (*A. vesiculosa*) develops snap-traps that are approximately 5 mm in length, and that can capture small aquatic prey such as water fleas (Figure 1c) [40]. The trap consists of two convex (seen from the outside) lobes that are connected by a midrib. Once prey triggers small sensitive hairs on the inner surfaces of the lobes, the trap closes within ca. 50 ms. Trap actuation involves turgor changes in motor cells and the release of prestress by downward bending (i.e., relaxation) of the prestressed midrib [40,41]. When the midrib flexes, the attached lobes are synchronously drawn together, and the trap shuts by a process called motion amplification [42]. This deformation behavior has been abstracted and implemented in the biomimetic façade shading Flectofold [43]. In the present work, we have attempted to develop artificial "traps" showing the complex coupled bending response involving motion amplification.

1.2.4. Blooming of the Lily Flower (*Lilium* spec.)

The opening of the lily flower (*L.* spec.) (Figure 1d) incorporates the bursting of the bud and, subsequently, the outward curving of the petals. Instead of differential growth processes at the adaxial (=upper) and abaxial (=lower) surfaces, the actuation principle of the individual petal is based on growth-induced cell elongation processes on the petal edges (and not on the central part), finally leading to wrinkled petal edges [44]. Our objective has been to transfer this edge-actuation into abstracted artificial paper-based "petals" that are capable of curvature changes similar to those of the natural petals.

1.2.5. Opening of the Legume Seed Pod

The chiral opening of the legume fruits (legume or seed pod) of many members of the Fabaceae (Figure 1e, exemplarily shown by *Spartium junceum*) is driven by shrinking processes of hygroscopic tissues. The initially straight valves of the fruit dry out, the occurring stresses lead to the rapid opening (popping) of the whole seed pod (thereby entailing ballistic seed release) [45], and the valves then slowly curl into helical strips of opposite handedness [46]. The structural bases for this flat-to-helical transition are the two fibrous layers that are found within each pod valve and that are oriented at ca. 45° with respect to the longitudinal axis of the pod. Upon drying out, the layers induce the observed deformation. The aim of this part of the project has been to transfer the basic functional morphology of the seed pod into a paper model incorporating the initial popping of the pod and the subsequent valve curling.

1.2.6. Snap-Trap of the Carnivorous Venus Flytrap (*Dionaea muscipula*)

The carnivorous Venus flytrap (*D. muscipula*) develops aerial traps of ca. 2 cm in length for the capture of arthropod prey. The triggering of sensitive hairs on the internal surface of the trap by prey causes the fast shutting of the trap within 100–500 ms. The snapping process is driven by turgor changes, the release of prestress, and a sudden concave-convex geometrical change of the individual lobes [3,4]. The goal here has been to develop a compliant system capable of hygroscopically induced snap-through behavior, thereby transferring the "speed-boosting" snap-through transition of the natural snap-trap.

2. Materials and Methods

We chose desiccation-driven motion actuation in all of our models. This type of actuation is, from a practical viewpoint, more readily achievable (compared with a water uptake and swelling-based mechanism) and offers a broader design space since the resulting constructions do not have to be placed into humid environments (e.g., water basins or chambers with high relative humidity) for the induction of swelling and movement. The motions of all biological concept generators were abstracted and transferred into paper-based models, except for the Venus flytrap (*D. muscipula*) model, for which a paper-polymer sheet combination was used. For the snap-trap of *A. vesiculosa*, a wood-polymer-based constructional approach was chosen additionally to the paper-based model. The models were photographed, and their motions were recorded with a Panasonic Lumix DMC-FZ 1000 bridge camera.

2.1. Paper-Based Models

We used "PERGA pastel" paper (weight: 100 g/m^2; Artoz Papier AG, Lenzburg, Germany) for the actuating layers (AL) and the construction paper "Tonpapier Weiß" (weight: 130 g/m^2; L. Jansen GmbH & Co. KG, Mönchengladbach, Germany) for the resistance layers (RL). We cut both types of paper (AL and RL) with scissors according to the schemes presented in the following.

The fiber orientations of each type of paper dictate the movement responses of the envisaged models and can be perceived macroscopically or with a magnifying glass. When the paper is torn, a smooth tear is produced in the fiber orientation of the paper, whereas the tear against the fiber direction is frayed and ragged.

For construction, we initially put the ALs in a water tank for approximately 5 min until they were fully soaked. Simultaneously, we coated the RLs with a thin film of UHU Max Repair glue (UHU, Bühl, Germany) at the respective connection areas, as indicated in the schemes presented in the following. We then carefully connected the two layers and placed them in a dry environment. The ALs shrank upon desiccation, which, in combination with the resistive properties of the RLs, dictated the deformation and movements of the models.

2.1.1. Flower of the Bird of Paradise (*S. reginae*)

The torsional buckling principle, as present in the flower sheath of *S. reginae*, can be transferred into a paper model via the construction steps depicted in Figure 2. A rectangular piece of paper forming the RL with a central slit and lengthwise fiber orientation, two paper flaps of any fiber orientation with folded edges (i.e., copy paper), and a rectangular wet AL with a transverse fiber arrangement are required. The folded edges of the flaps are inserted through the slit in the RL and glued onto its lower surface. The wet AL is then also glued onto this lower surface. The whole model is then set up in a dry environment, with the flaps oriented upwards, onto a narrow structure allowing the bending of the AL/RL bilayer.

2.1.2. False Rose of Jericho (*S. lepidophylla*)

The composite architecture as present in the natural *S. lepidophylla* can be abstracted and transferred into a paper model via the construction steps depicted in Figure 3. First, a small circular middle gluing point is cut out of paper, regardless of which type of fiber orientation. Subsequently, numerous individual bilayer "stems" are produced that consist of dry RLs with lengthwise fiber orientations glued to identically shaped wet ALs with transverse fiber orientations. The "stems" are finally glued to the middle point.

2.1.3. Snap-Trap of the Carnivorous Waterwheel Plant (*A. vesiculosa*)

Curved-fold bending, as present in the natural trap of *A. vesiculosa*, can be abstracted and transferred into a paper model via the construction steps depicted in Figure 4. First, a circular piece of paper is folded in such a way that an elliptical middle lens is created, representing the midrib of the natural trap and constituting the RL in our model. The fiber orientation of the paper should lie along the length of this lens. The wet AL, which is of the identical elliptical shape as the RL (lens) but with transverse fiber orientation, is glued onto the underside. The remaining parts of the round paper are folded upwards along the crease lines and, thereby, become the convex (as seen from the outside) lateral flaps. During this folding procedure, the lens becomes slightly arched.

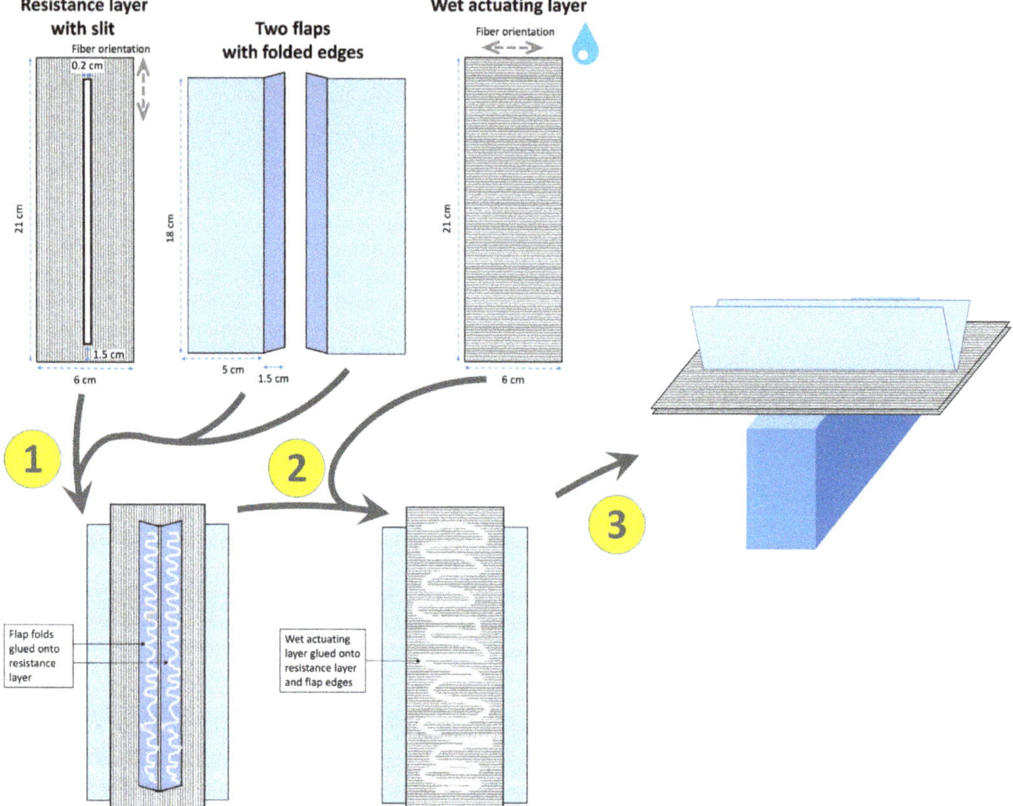

Figure 2. Exemplary construction scheme for the abstraction of the torsional buckling-induced bending deformation of the flower sheath of the bird of paradise (*S. reginae*) and its transfer into a paper-based self-actuated compliant system. The various construction steps are numbered (1, 2, 3). The fiber orientations of the dry RL and wet AL are indicated (gray dashed arrows). Details are provided in the main text.

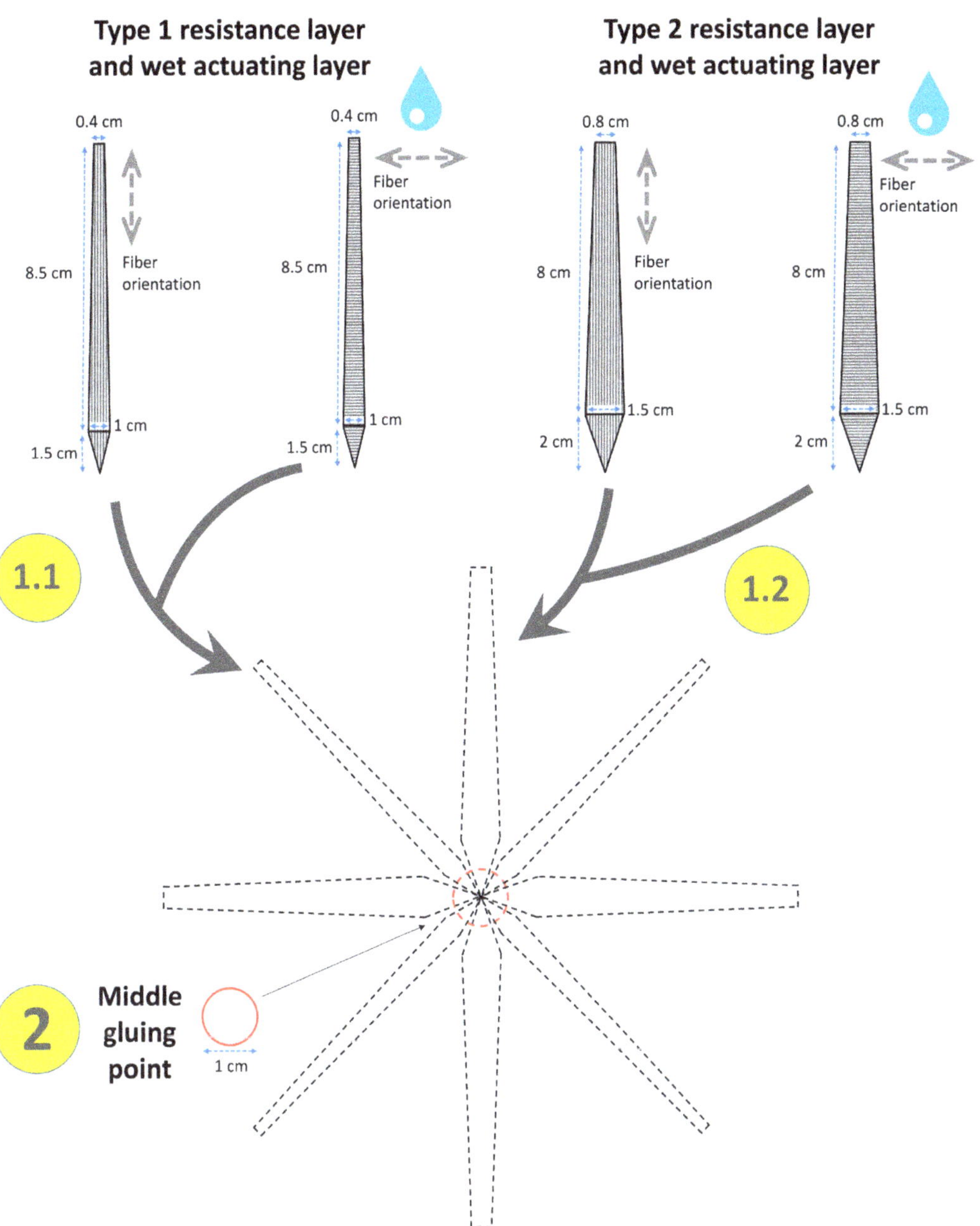

Figure 3. Exemplary construction scheme for the abstraction of the composite structure of the spike-moss *S. lepidophylla* and its transfer into a paper-based self-actuated compliant system. The various construction steps are numbered (1, 2). Steps 1.1 and 1.2 (construction of "stems" and their attachment by glue to the circular middle point) do not have to be undertaken in any particular order and can be repeated as many times as required (in our case: four each). The fiber orientations of the dry RLs and wet ALs are indicated (gray dashed arrows). Details are provided in the main text.

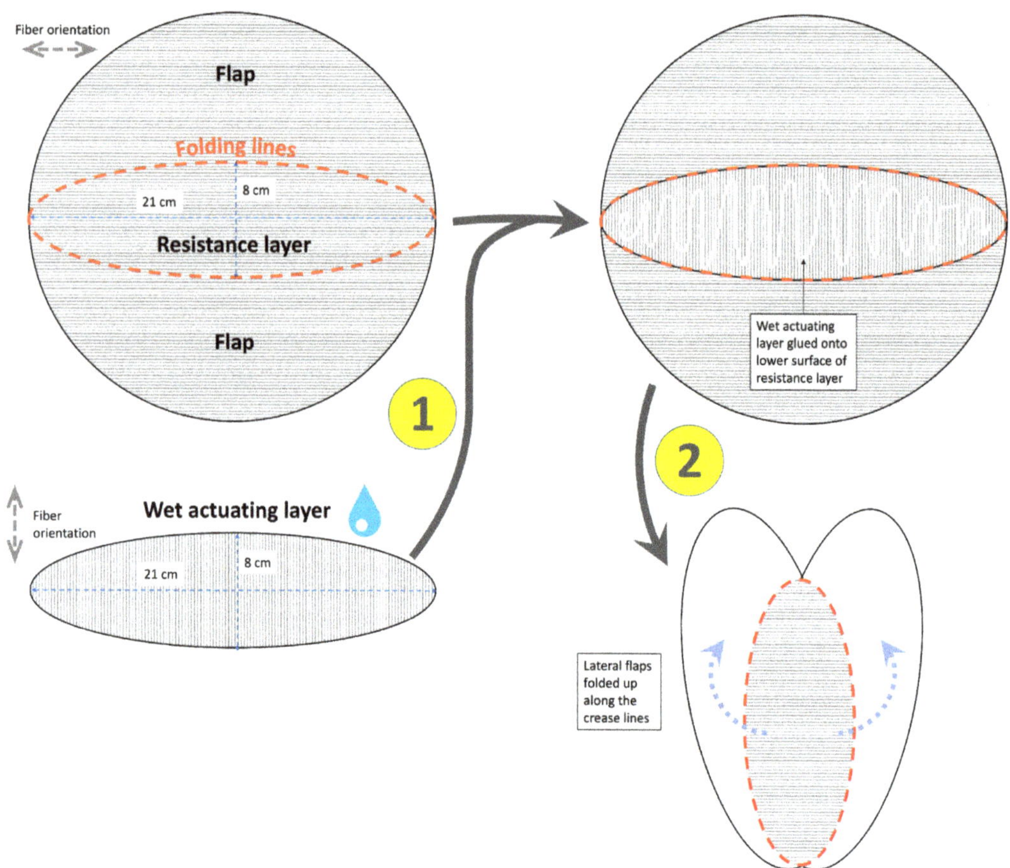

Figure 4. Exemplary construction scheme for the abstraction of the *A. vesiculosa* snap-trap and its transfer into a paper-based self-actuated compliant system. The various construction steps are numbered (1, 2). The fiber orientations of the dry RL and wet AL are indicated (gray dashed arrows). Details are provided in the main text.

2.1.4. Blooming of the Lily Flower (*L.* Spec.)

The petal deformation principle of the lily (*L.* spec.) can be abstracted and transferred into a paper model via the construction steps depicted in Figure 5. First, an elliptical RL is cut out having a lengthwise fiber orientation. For the AL, another ellipse of identical shape but with a transverse fiber orientation is created and then cut in such a way that only the upper margin remains. This margin is then glued onto the RL, thereby creating a bilayer zone. For convenience, we also constructed a holder made from cardboard and adhesive tape.

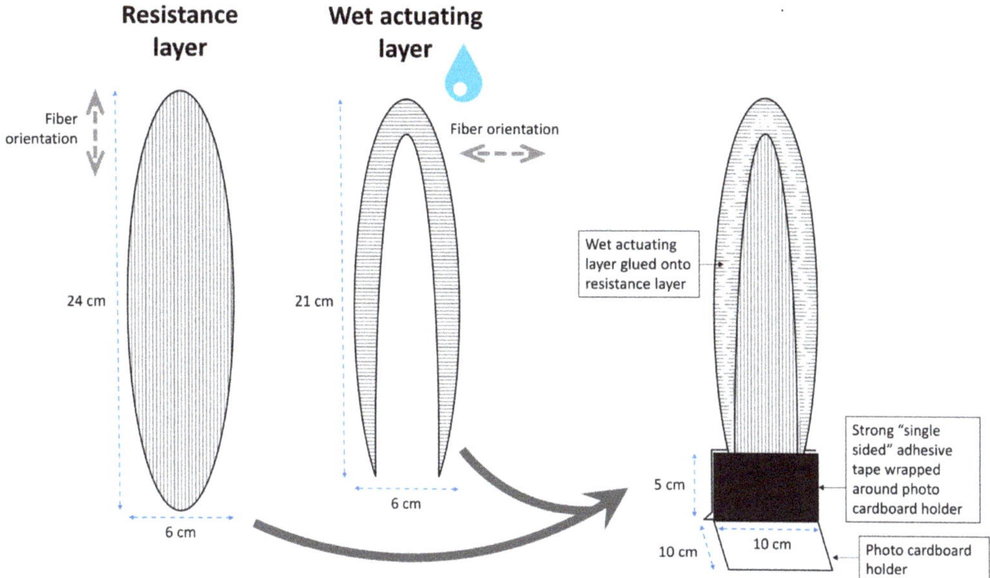

Figure 5. Exemplary construction manual for the abstraction of the deformation of the petal of the lily (*L.* spec.) and its transfer into a paper-based self-actuated compliant system. The fiber orientations of the dry RL and wet AL are indicated (gray dashed arrows). Details are provided in the main text.

2.1.5. Opening of the Legume

Legume opening and valve twisting can be abstracted and transferred into a paper model via the construction steps depicted in Figure 6. The two valves are constructed as bilayers: the RLs both possess lengthwise fiber orientations, whereas the two wet ALs possess opposite fiber orientations at 45° with respect to their longitudinal axes. Both initially straight valves are then connected at defined areas with double-sided adhesive tape (3M 928 Transferklebeband).

2.2. Paper-Polymer Model

For the transfer of the snap-buckling instability of the Venus flytrap (*D. muscipula*), we used the same wet AL paper and glue as indicated in Section 2.1. The polymer sheet was taken by cutting the opaque thicker side of a clip-folder.

Snap-Trap of the Carnivorous Venus Flytrap (*D. muscipula*)

The snap-through instability, which speed boosts the trap closure in the Venus flytrap (*D. muscipula*), can be transferred into a paper-polymer model (Figure 7). A strip taken from the polymer sheet acts as the RL and performs the striking curvature inversion, which is initiated by the shrinking processes of two initially wet ALs. The wet ALs are glued on the strip as depicted in Figure 7 so that their fiber orientations are transverse to the longitudinal axis of the strip.

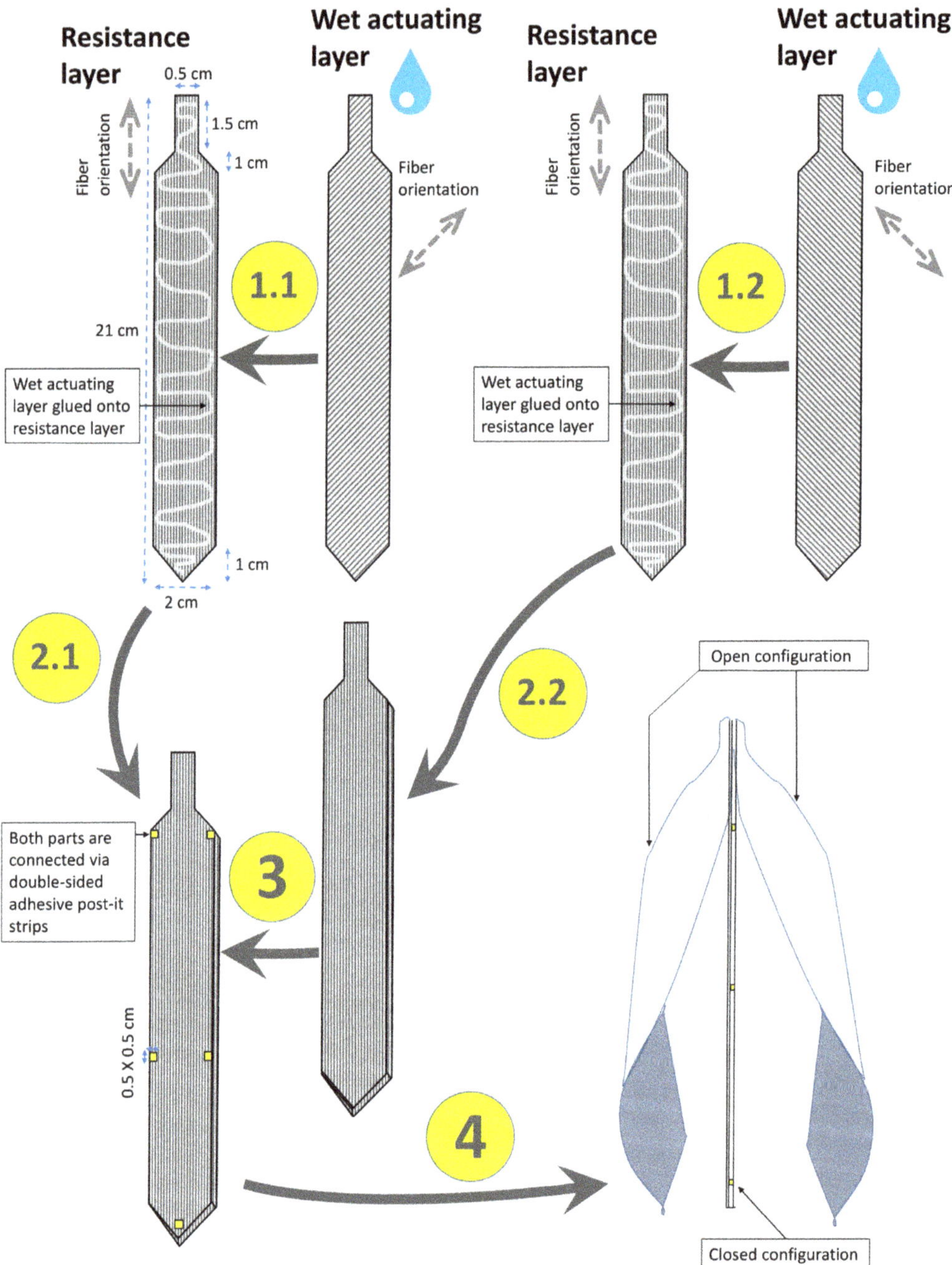

Figure 6. Exemplary construction scheme for the abstraction of the legume (Fabaceae seed pod) and its transfer into a paper-based self-actuated compliant system. The fiber orientations of the dry RLs and wet ALs are indicated (gray dashed arrows). The various construction steps are numbered (1–4). Steps 1.1/1.2 and 2.1/2.2, respectively, can be completed in any order. Details are provided in the main text.

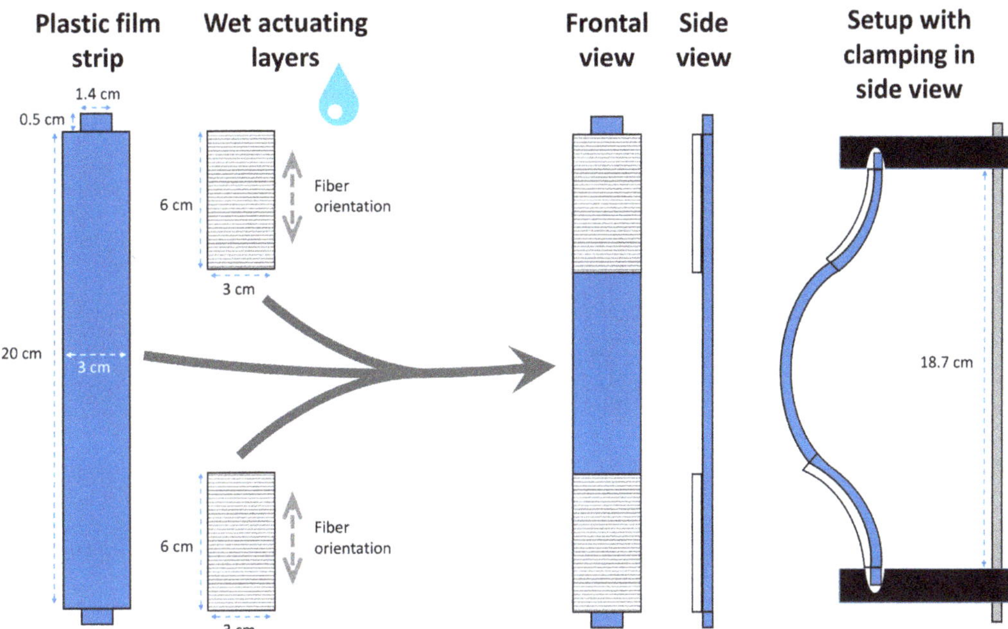

Figure 7. Exemplary construction scheme for the abstraction of snap-through instability as present in the Venus flytrap (*D. muscipula*) and its transfer into a paper-based self-actuated compliant system. The fiber orientations of the wet ALs are indicated (gray dashed arrows). Details are provided in the main text.

2.3. Wood-Polymer Models

A similar desiccation-driven motion actuation involving a coupled bending system has been chosen for the wood-polymer structures that mimic the motion of *A. vesiculosa*. The materials needed for the wood-polymer structure ("Flectofold-like") are: household string, a clip-folder, and five wooden veneer strips. Six radii of curvature are used as in [43]. The wood-polymer structure is composed of two major substructures: a multilayered backbone composite (wood) and a foldable sheet (polymer). The substructures are manufactured separately and then assembled.

Snap-Trap of the Carnivorous Waterwheel Plant (*A. vesiculosa*)

To produce an actuating multilayered backbone composite, five veneer strips with the dimensions 20 × 190 × 0.5 mm are cut out of a larger sheet of veneer by using a paper cutter. Four veneer strips are cut perpendicular to the grain (actuating zone), and one is cut parallel to the grain (resistance zone). The strips are immersed in water (for approximately 20 s, completely submerged) and are then wiped with a cloth to remove excess water. In the next step, the strips are glued to each other by using water-proof wood glue (Ponal Wasserfest, Henkel, Germany) so as to form a five-layered composite backbone, with the strip cut parallel to the grain being placed on the outside of the composite backbone. The backbone is then clamped between two wooden boards that are slightly larger than the backbone. This ensures that the backbone remains straight during clamping. To prevent the wooden boards from sticking to the wooden strips, a polymer sheet (transparent side of the clip-folder) acts as a "cover" for the boards so that the boards are not in direct contact with the wooden strips (Figure 8). The clamps are then attached to the wooden boards, and the structure is left to set for two hours at room temperature.

Figure 8. Various steps for clamping the veneer composite by means of boards and clamps in a plastic sleeve.

The foldable polymer sheet is cut out using a template (Figure 9). This template is printed on paper, cut out, and placed on the rear "hard" side of a clip-folder (Durable Dura-clip 30). The rear side is then cut into the same shape as the paper template. One of the six provided radii of curvature (150, 170, 200, 250, 330, 500 mm) on the template can now be chosen. To transfer the curvature from the paper template to the clip-folder sheet, the points on the line of the paper template are pierced with a needle through to the polymer sheet. This creates a hole pattern along the selected curvature lines on the polymer sheet. Furthermore, 10 holes marked on the paper template (to the right and left of the vertical axis) are punched through the polymer sheet by using a 4 mm punching iron. The paper template is then removed, and the clip-folder sheet is folded along the curved lines. The polymer sheet must be folded back and forth from one side to the other about 30 times to achieve certain flexibility along the folding curvature.

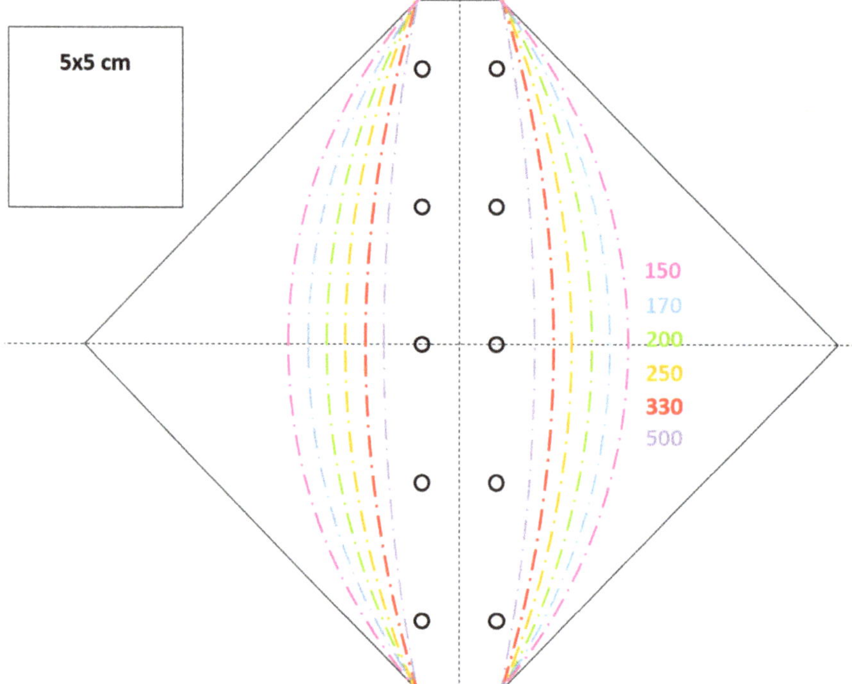

Figure 9. Printable template with six radii of curvature.

Next, the wood-polymer structure is assembled (Figure 10). The clamps are removed from the multilayered veneer backbone, which is then fixed together with the polymer sheet by threading each of five pieces of household string (approximately 20 cm in length) through the pairs of large holes in the clip-folder sheet and knotting each piece of string. The following steps are of importance: (1) the strip cut parallel to the grain is on top, and (2) the triangular side-pieces of the polymer sheet are slightly bent upwards to induce a folding direction.

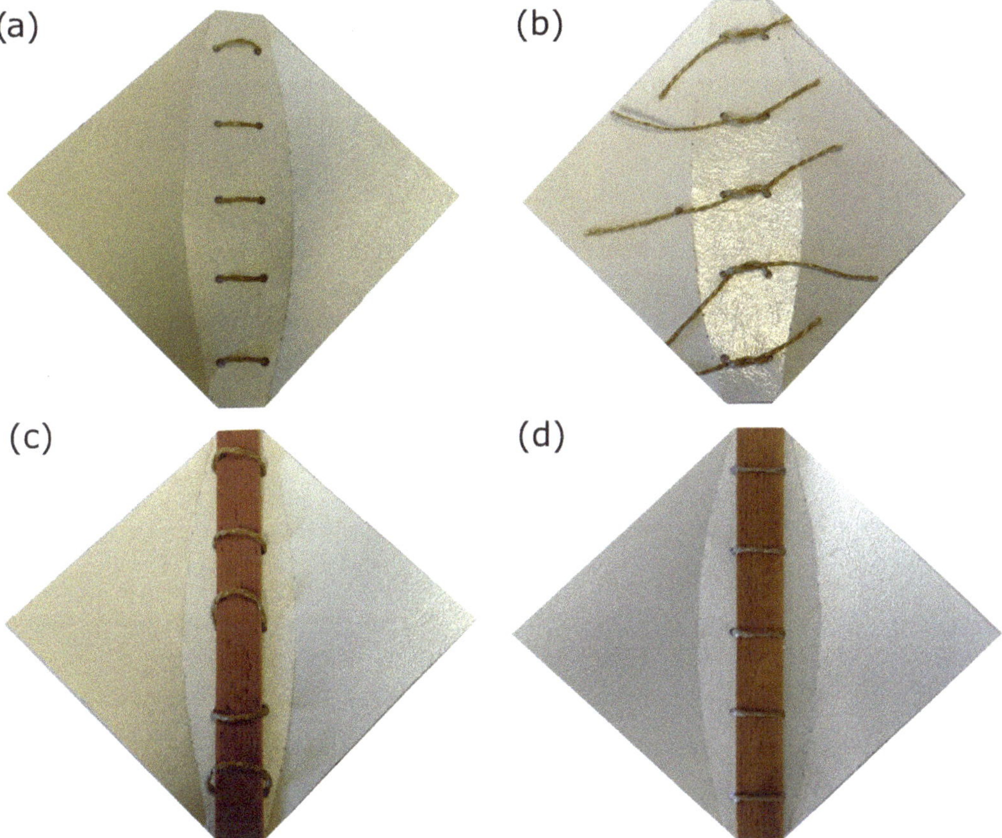

Figure 10. Various steps for connecting the Flectofold-like demonstrator to the actuating multilayer veneer composite. The five pieces of string are threaded through the adjacent large holes in the polymer sheet (**a**), and each is loosely knotted on the back of the demonstrator by an overhand knot (**b**). The wooden veneer is pushed through the string loops (**c**) and firmly attached to the polymer sheet (**d**) by tightening the overhand knots and then multiply knotting the strings with overhand knots on the back of the demonstrator.

3. Results

In the following, we present examples of solutions for the biomimetic transfer of actuation and/or deformation processes of interest. The geometries and dimensions indicated in the construction "schemes" and the paper and glue types indicated in the Materials and Methods section are not mandatory. Indeed, many models can be geometrically distorted or constructed with other materials (e.g., other paper and glue types) and still show the anticipated movements. However, the presented models all have in common that they function reliably, are easy to assemble, and are small in size. Additionally, the individual materials of the wood-polymer models can be easily separated from each other and disposed of.

3.1. Paper-Based Models

3.1.1. Flower of the Bird of Paradise (*S. reginae*)

In our tested model (Figure 11, Video S1), the whole drying-induced bending process lasted approximately 22 min. The initially upwards oriented flaps gradually bend laterally downward, as dictated by the increasing curvature of the AR/AL bilayer. Thereby, the whole structure opens out, similar to the natural flower sheath, which opens under the weight force of the pollinating bird.

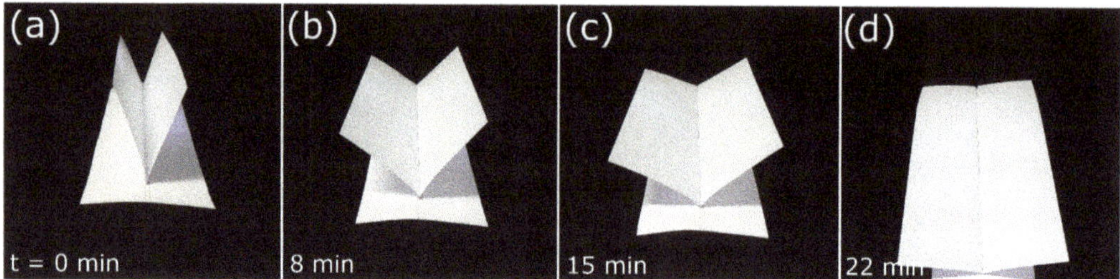

Figure 11. A self-actuated paper-based compliant system inspired by the bird of paradise flower (*S. reginae*). (**a**–**d**) Opening sequence during desiccation. Time is indicated bottom left.

3.1.2. False Rose of Jericho (*S. lepidophylla*)

We chose two types of "stems" differing in size, with type 1 being smaller than type 2, in our model. Each type was constructed four times. Upon drying, the compliant system showed a striking composite closure motion, with type 1 "stems" curling faster than type 2 (Figure 12, Video S2). The overall motion was completed after 122 min, and a closed state similar to the natural counterpart (cf. Figure 1b) was attained.

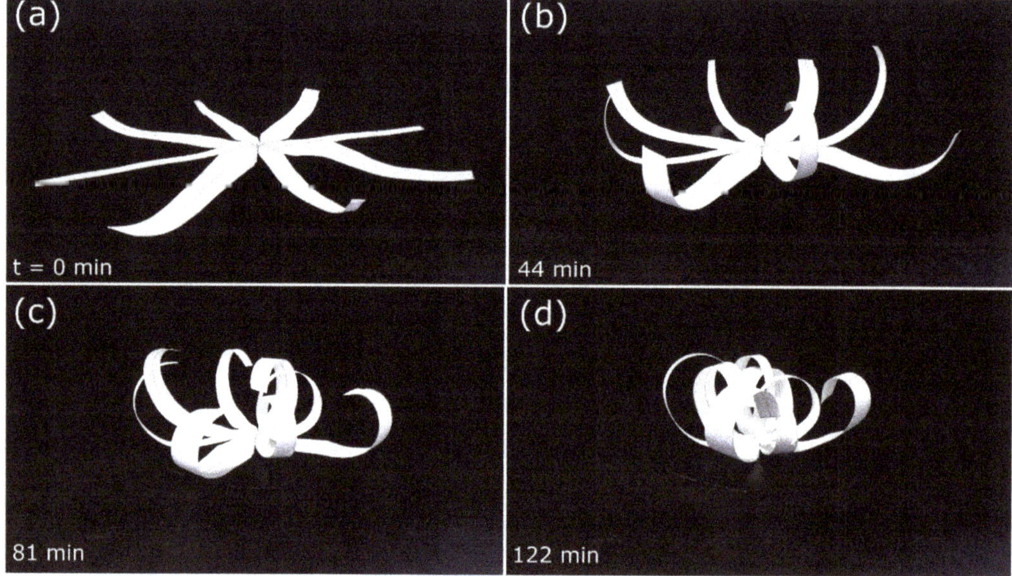

Figure 12. Self-actuated paper-based compliant system inspired by the composite architecture of *S. lepidophylla* performing closure upon drying. (**a**–**d**) Folding sequence during desiccation. Time is indicated bottom left.

3.1.3. Snap-Trap of the Carnivorous Waterwheel Plant (*A. vesiculosa*)

Full closure of the artificial "trap" in our model occurred within 72 min (Figure 13, Video S3). The lens (rib) progressively bent during the drying process, and the lateral folds were drawn together. In the closed state, the folds (representing the natural trap lobes) pressed against each other.

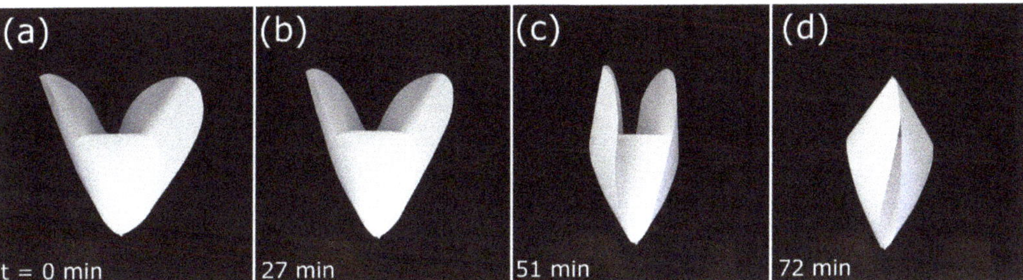

Figure 13. Self-actuated paper-based compliant system inspired by the snap-trap of the waterwheel plant (*A. vesiculosa*) performing closure upon drying. (**a–d**) Snapping sequence during desiccation. Times are indicated bottom left.

3.1.4. Blooming of the Lily Flower (*L. spec.*)

The model shows a striking bending deformation during drying, which is completed after 42 min (Figure 14, Video S4) and is highly reminiscent of the blooming motion of the natural petal.

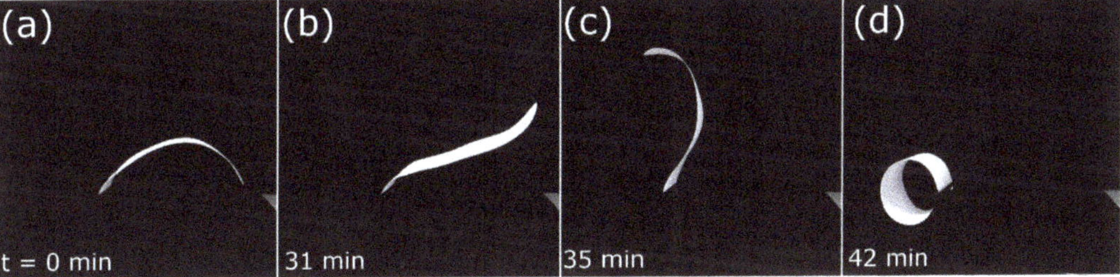

Figure 14. Self-actuated paper-based compliant system inspired by the lily petal (*L. spec.*). (**a–d**) Bending sequence during desiccation. Times are indicated bottom left.

3.1.5. Opening of the Legume

The drying-induced deformation of the valves in our model induced their rapid splitting (popping) after 33 min (Figure 15, Video S5). In the natural seed pod, this splitting leads to the scattering of the seed. The opposite-handed twisting of the two valves continues until maximum seed pod opening is achieved after 52 min.

3.2. Paper-Polymer Model

Snap-Trap of the Carnivorous Venus Flytrap (*D. muscipula*)

We used a clamping system to install the initial "outward" curvature of the polymer sheet strip with the two ALs (Figure 16, Video S6) in our model. The drying-induced shrinking of the ALs caused a continuous deformation of the strip until its curvature suddenly flipped "inward" at $t = 29$ min.

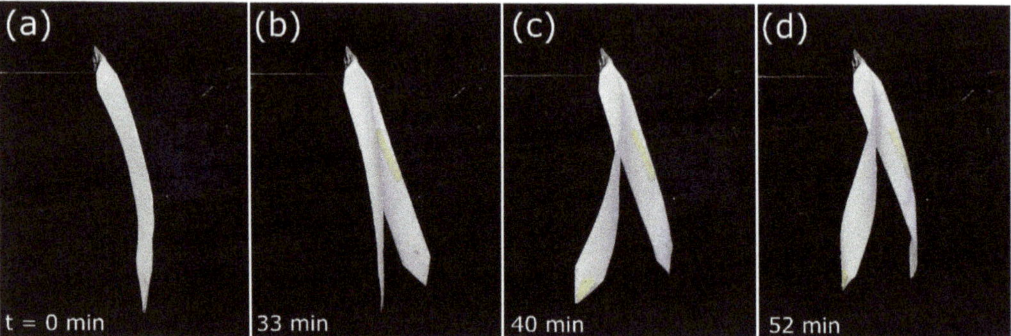

Figure 15. Self-actuated paper-based compliant system inspired by the legume (Fabaceae seed pod) performing rapid opening (at *t* = 33 min) (**a**,**b**) and continuous opposite-handed twisting of the valves during desiccation (**b**–**d**). Times are indicated bottom left.

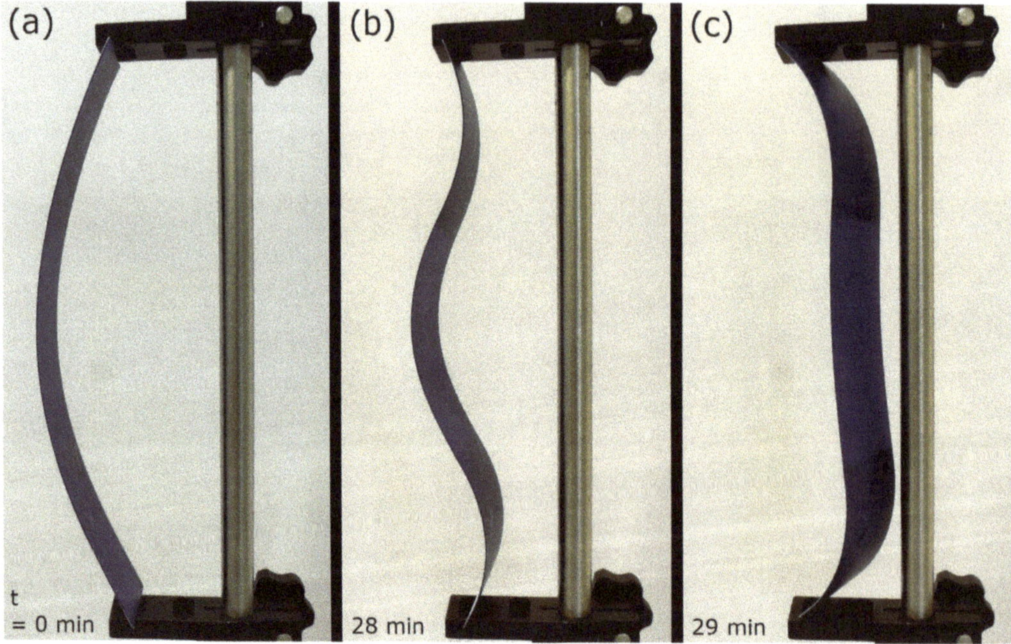

Figure 16. Self-actuated paper-polymer-based compliant system inspired by the Venus flytrap (*D. muscipula*) performing a rapid snap-through transition (at *t* = 29 min). (**a**) Initial state. (**b**) Half-completed movement, shortly before snap-buckling. (**c**) Final state. Times are indicated bottom left.

3.3. Wood-Polymer Models

Snap-Trap of the Carnivorous Waterwheel Plant (*A. vesiculosa*)

The bending and folding process of the polymer sheet attributable to the desiccation of the five-layered composite backbone is completed after approximately 24–48 h. Even though the required actuation force cannot be measured, the relationship between the radius of curvature of the fold, the radius of curvature of the wood backbone, and the maximal closing angle during the closing of the Flectofold-like structures can be observed (Figure 17). The closing motion of two structures with a radius of curvature of 200 and

250 mm can be followed in the Supplementary Materials (Video S7) and for two structures with a radius of curvature of 330 and 500 mm (Video S8).

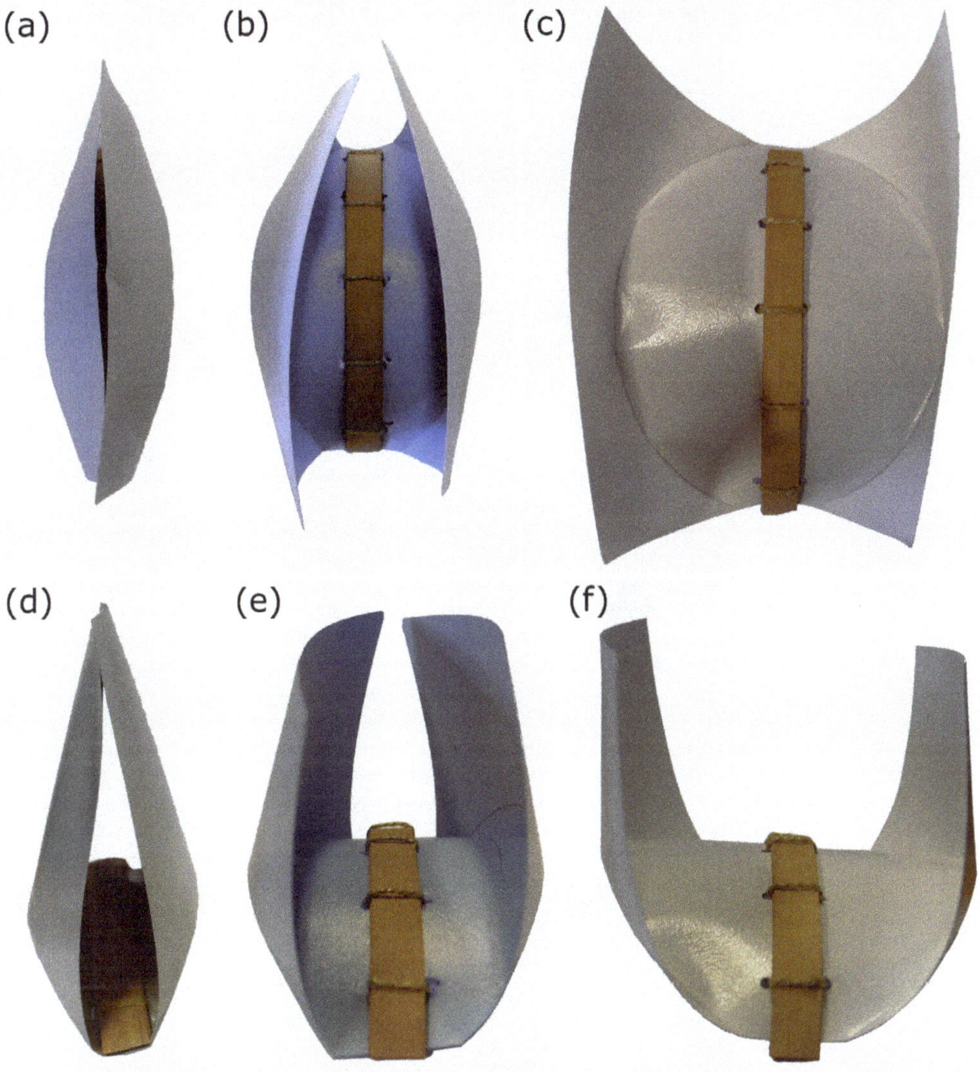

Figure 17. Top and side view of various Flectofold-like structures after the completed movement. The positive correlation between the radius of curvature and the maximal closing angle of the "wings" can be easily discerned. Radius of curvature: (**a**,**d**) 190 mm, (**b**,**e**) 150 mm, and (**c**,**f**) 120 mm.

4. Discussion

The abstraction of plant movement principles and their transfer into biomimetic 4D printed structures form part of a rapidly progressing research field [10–16,18–28]. However, 4D printing requires special user skills and expensive hardware hindering its application in research and teaching, which cannot cover such high costs. Our approach offers a low-cost alternate methodology for developing biomimetic self-actuated compliant systems for applications, e.g., in the context of didactic approaches, arts-crafting, and sensors (i.e., humidity). With regard to material costs, each of the paper-based models presented in this paper is in the sub-Euro (€) region and, therefore, affordable for everyone. The models that include polymer foil from clip-holders are only slightly more expensive (in the low €-region). Although the use of paper and wood for handcrafting hygroscopic structures is not new [33], the transfer and successful technical implementation of complex plant deformation principles as presented here have not been previously achieved. We are convinced that many more plant movement and deformation principles can be similarly abstracted and produced. Promising potential concept generators in which the above-described concept can be applied include the hygroscopic seed capsules of the ice plant (*Delosperma nakurense*) [47], the twisting and hook-forming root hairs of English ivy (*Hedera helix*) [48], and the coiling of the awn of the stork's bill (*Erodium gruinum*) [49].

Most of the hygroscopically actuated models presented in this paper perform desiccation-induced and, thereby, shrinkage-driven movements. Exceptions are the models of the legume (Fabaceae seed pod) and of the snap-trap of *D. muscipula* in which the motion additionally incorporates elastic instabilities. Thus, complex combinations of various movement actuation principles can indeed be achieved with simple manual construction methods.

With regard to the lily petal structure (Figures 5 and 14, Video S4), we must state that, in the natural flower, expansion (growth) of the outer petal zone (edge) takes place, which is accompanied by petal wrinkling. However, despite numerous attempts, we have not been able to create a similar marginal zone on the artificial "petal". Therefore, the marginal zone in our model, which is constructed as a functional bilayer, represents an additional abstraction step leading to the observed motion.

Whereas several concept generators used in this study show repeatable motions (*S. reginae* flower, *S. lepidophylla* stems, snap-traps of *A. vesiculosa* and *D. muscipula*), some other processes (slow *Lilium* flower opening, rapid Fabaceae seed pod opening) are not repeatable, and the respective structures, thereby, constitute "single-use" devices (in the context of the performed motions). Due to structural disintegration during selective rewetting of the actuating layers, the presented paper-based models also function only once, whereas the paper-polymer model (at least to some extent) and, specifically, the wood-polymer models can be actuated numerous times. The Flectofold-like structures (Figure 17) can be examined during the process of reopening during drying. Therefore, such more expensive and elaborate models have a reversibility feature as an extra.

Our models were not developed to act as "product demonstrators" or similar milestones in research and development approaches. This can only be achieved by precise additive manufacturing and other reliable methods. Instead, our models should be regarded as (1) easy-to-achieve functional demonstrators for the proof-of-concept of certain functionalities (e.g., the interplay of structure and motion as revealed here), and (2) as a means for teaching the "manual comprehension" of plant structure, mechanics, and movement. In particular, the low constructional costs and the easy and time-saving manufacturing procedures allow the production of numerous physical models and model generations. In the context of teaching, students are able to develop technical counterparts of motile plant structures and to demonstrate physically correct (or abstracted) movement actuation and/or deformation processes by such modeling. The involved processes of abstracting biological working principles and their transfer into technical materials require a wealth of ideas and spatial imagination on the part of the students who can either follow a set of instructions (such as the construction schemes presented here) or their own inspiration. This has an enormously motivating effect on learning and creates a deeper understanding of fundamental mechanical principles both in a plant-inspired technical application (biomimetics) and in the biological model (reverse biomimetics).

Supplementary Materials: The following are available online at https://www.mdpi.com/2313-767 3/6/3/42/s1, Video S1: A self-actuated paper-based compliant system inspired by the bird of paradise flower (*S. reginae*), playback rate: 10 fps (100× time lapse), Video S2: A self-actuated paper-based compliant system inspired by the composite architecture of *S. lepidophylla*, playback rate: 10 fps (100× time lapse), Video S3: A self-actuated paper-based compliant system inspired by the snap-trap of the waterwheel plant (*A. vesiculosa*), playback rate: 10 fps (100× time lapse), Video S4: A self-actuated paper-based compliant system inspired by the lily petal (*L.* spec.), playback rate: 10 fps (100× time lapse), Video S5: A self-actuated paper-based compliant system inspired by the legume fruit (Fabaceae seed pod), playback rate: 10 fps (100× time lapse), Video S6: A self-actuated paper-polymer-based compliant system inspired by the Venus flytrap (*D. muscipula*), playback rate: 10 fps (100× time lapse), Video S7: Closing motion of a wood-polymer-based Flectofold structure with 200 und 250 mm radius of curvature, playback rate: 10 fps (2400× time lapse), Video S8: Closing motion of a wood-polymer-based Flectofold structure with 330 und 500 mm radius of curvature, playback rate: 10 fps (2400× time lapse). Videos with higher resolution can be obtained from the authors upon request.

Author Contributions: Conceptualization, S.P., O.S., T.S. and T.M.; Formal analysis, S.P., P.S. and T.M.; Investigation, S.P., P.S., O.S., T.S., B.B. and T.M.; Methodology, S.P., P.S. and T.M.; Writing—original draft, S.P. and T.M.; Writing—review and editing, O.S., T.S. and B.B. All authors have read and agreed to the published version of the manuscript.

Funding: S.P. and T.S. gratefully acknowledge BASF SE, Ludwigshafen, Germany, and the Ministry of Science Research and Arts of the State of Baden-Württemberg, Germany, who financially supported this research within the frame of the Academic Research Alliance JONAS ("Joint Research Network on Advanced Materials and Systems") established jointly with BASF SE and the University of Freiburg, Germany. T.S., O.S. and S.P. further acknowledge funding by the Deutsche Forschungsgemeinschaft (DFG, German Research Foundation) under Germany's Excellence Strategy—EXC-2193/1—390951807, and by the collaborative project "Bio-inspirierte elastische Materialsysteme und Verbundkomponenten für nachhaltiges Bauen im 21ten Jahrhundert" (BioElast) within the "Zukunftsoffensive IV Innovation und Exzellenz—Aufbau und Stärkung der Forschungsinfrastruktur im Bereich der Mikro- und Nanotechnologie sowie der neuen Materialien" of the State Ministry of Baden-Wuerttemberg for Sciences, Research and Arts. The article processing charge was funded by the Baden-Wuerttemberg Ministry of Science, Research and Art and the University of Freiburg in the funding program Open Access Publishing.

Institutional Review Board Statement: Not applicable.

Informed Consent Statement: Not applicable.

Data Availability Statement: All relevant data are included within the paper and its Supplementary Materials files.

Conflicts of Interest: The authors declare no conflict of interest.

References

1. Skotheim, J.; Mahadevan, L. Physical limits and design principles for plant and fungal movements. *Science* **2005**, *308*, 1308–1310. [CrossRef]
2. Dumais, J.; Forterre, Y. "Vegetable dynamicks": The role of water in plant movements. *Annu. Rev. Fluid Mech.* **2012**, *44*, 453–478. [CrossRef]
3. Forterre, Y. Slow, fast and furious: Understanding the physics of plant movements. *J. Exp. Bot.* **2013**, *64*, 4745–4760. [CrossRef]
4. Poppinga, S.; Masselter, T.; Speck, T. Faster than their prey: New insights into the rapid movements of active carnivorous plants traps. *BioEssays* **2013**, *35*, 649–657. [CrossRef]
5. Weintraub, M. Leaf movements in *Mimosa pudica* L. *N. Phytol.* **1952**, *50*, 357–382. [CrossRef]
6. Dawson, C.; Vincent, J.F.; Rocca, A.M. How pine cones open. *Nature* **1997**, *390*, 668. [CrossRef]
7. Forterre, Y.; Skotheim, J.M.; Dumais, J.; Mahadevan, L. How the *Venus flytrap* snaps. *Nature* **2005**, *433*, 421–425. [CrossRef]
8. Sachse, R.; Westermeier, A.; Mylo, M.; Nadasdi, J.; Bischoff, M.; Speck, T.; Poppinga, S. Snapping mechanics of the Venus flytrap (*Dionaea muscipula*). *Proc. Natl. Acad. Sci. USA* **2020**, *117*, 16035–16042. [CrossRef] [PubMed]
9. Howell, L.L. *Compliant Mechanisms*; Wiley: New York, NY, USA, 2001.
10. Poppinga, S.; Körner, A.; Sachse, R.; Born, L.; Westermeier, A.S.; Hesse, L.; Knippers, J.; Bischoff, M.; Gresser, G.T.; Speck, T. Compliant mechanisms in plants and architecture. In *Biomimetic Research for Architecture and Building Construction: Biological Design and Integrative Structures*; Knippers, J., Speck, T., Nickel, K., Eds.; Springer: Berlin/Heidelberg, Germany, 2016; pp. 169–193.

11. Burgert, I.; Fratzl, P. Actuation systems in plants as prototypes for bioinspired devices. *Philos. Trans. R. Soc. A* **2009**, *28*, 1541–1557. [CrossRef] [PubMed]

12. Schleicher, S.; Lienhard, J.; Poppinga, S.; Speck, T.; Knippers, J. A methodology for transferring principles of plant movements to elastic systems in architecture. *Comput. Aided Des.* **2015**, *60*, 105–117. [CrossRef]

13. Li, S.; Wang, K.W. Plant-inspired adaptive structures and materials for morphing and actuation: A review. *Bioinsp. Biomim.* **2017**, *12*, 011001. [CrossRef] [PubMed]

14. Mathijsen, D. The rise of compliant mechanisms: composites' role in enabling an engineering revolution. *Reinf. Plast.* **2020**, *64*, 254–260. [CrossRef]

15. Guo, Q.; Dai, E.; Han, X.; Xie, S.; Chao, E.; Chen, Z. Fast nastic motion of plants and bioinspired structures. *J. R. Soc. Interface* **2015**, *12*, 20150598. [CrossRef] [PubMed]

16. Quian, X.; Zhao, Y.; Alsaid, Y.; Wang, X.; Hua, M.; Galy, T.; Gopalakrishna, H.; Yang, Y.; Cui, J.; Liu, N.; et al. Artificial phototropism for omnidirectional tracking and harvesting of light. *Nat. Nanotechnol.* **2019**, *14*, 1048–1055. [CrossRef]

17. Harlow, W.M.; Côté, W.A.; Day, A.C. The opening mechanism of pine cone scales. *J. For.* **1964**, *62*, 538–540. [CrossRef]

18. Reyssat, E.; Mahadevan, L. Hygromorphs: From pine cones to biomimetic bilayers. *J. R. Soc. Interface* **2009**, *6*, 951–957. [CrossRef]

19. Ionov, L. Biomimetic hydrogel-based actuating systems. *Adv. Funct. Mater.* **2013**, *23*, 4555–4570. [CrossRef]

20. Stoychev, G.; Guiducci, L.; Turcaud, S.; Dunlop, J.W.C.; Ioniv, L. Hole-programmed superfast multistep folding of hydrogel bilayers. *Adv. Funct. Mater.* **2016**, *26*, 7733–7739. [CrossRef]

21. Le Duigou, A.; Requile, S.; Beaugrand, J.; Scarpa, F.; Castro, M. Natural fibres actuators for smart bio-inspired hygromorph biocomposites. *Smart Mater. Struct.* **2017**, *26*, 125009. [CrossRef]

22. Erb, R.M.; Sander, J.S.; Grisch, R.; Studart, A.R. Self-shaping composites with programmable bioinspired microstructures. *Nat. Commun.* **2013**, *4*, 1712. [CrossRef]

23. Bargardi, F.L.; Le Ferrand, H.; Libanori, R.; Studart, A.R. Bio-inspired self-shaping ceramics. *Nat. Commun.* **2016**, *7*, 13912. [CrossRef]

24. Gladman, A.S.; Matsumoto, E.A.; Nuzzo, R.G.; Mahadevan, L. Biomimetic 4D printing. *Nat. Mater.* **2016**, *15*, 413–418. [CrossRef]

25. Correa, D.; Menges, A. Fused filament fabrication for multi-kinematic-state climate-responsive aperture. In *Fabricate 2017*; Menges, A., Sheil, B., Glynn, R., Eds.; UCL Press: London, UK, 2017; pp. 190–195.

26. Joshi, S.; Rawat, K.; Karunakaran, C.; Rajamohan, V.; Mathew, A.T.; Koziol, K.; Thakur, V.K.; Balan, A.S.S. 4D printing of materials for the future: Opportunities and challenges. *Appl. Mater. Today* **2020**, *18*, 100490. [CrossRef]

27. Correa, D.; Poppinga, S.; Mylo, M.D.; Westermeier, A.S.; Bruchmann, B.; Menges, A.; Speck, T. 4D pine scale: Biomimetic 4D printed autonomous scale and flap structures capable of multi-phase movement. *Philos. Trans. R. Soc. A* **2020**, *378*, 20190445. [CrossRef]

28. Poppinga, S.; Correa, D.; Bruchmann, B.; Menges, A.; Speck, T. Plant movements as concept generators for the development of biomimetic compliant mechanisms. *Integr. Comp. Biol.* **2020**, *60*, 886–895. [CrossRef] [PubMed]

29. Poppinga, S.; Lienhard, J.; Schleicher, S.; Speck, O.; Knippers, J.; Speck, T.; Masselter, T. Paradiesvogelblume trifft Architektur—Bionische Innovation für gelenkfreie technische Anwendungen. *Prax. Nat. Biol. Sch.* **2012**, *5*, 31–35.

30. Poppinga, S.; Metzger, A.; Speck, O.; Masselter, T.; Speck, T. Schnappen, schleudern, saugen: Fallenbewegungen fleischfressender Pflanzen. *Biol. Unserer Zeit* **2013**, *43*, 352–361. [CrossRef]

31. Salmén, L.; Boman, R.; Fellers, C.; Htun, M. The implications of fiber and sheet structure for the hygroexpansivity of paper. *Nord. Pulp. Paper. Res. J.* **1987**, *4*, 127–131. [CrossRef]

32. Skaar, C. *Wood-Water Relation*; Springer: Berlin/Heidelberg, Germany, 1988.

33. Brauner, L.; Rau, W. *Versuche zur Bewegungsphysiologie der Pflanzen*; Pflanzenphysiologische Praktika Band 3; Springer: Berlin/Heidelberg, Germany, 1966.

34. Rowan, M.K. Bird pollination of *Strelitzia*. *Ostrich* **1974**, *45*, 40.

35. Kronestedt, E.; Walles, B. Anatomy of the *Strelitzia reginae* flower (Strelitziaceae). *Nord. J. Bot.* **1986**, *6*, 307–320. [CrossRef]

36. Lienhard, J.; Schleicher, S.; Poppinga, S.; Masselter, T.; Milwich, M.; Speck, T.; Knippers, J. Flectofin: A hingeless flapping mechanism inspired by nature. *Bioinspir. Biomim.* **2011**, *6*, 045001. [CrossRef]

37. Poppinga, S.; Smaij, J.; Westermeier, A.S.; Horstmann, M.; Kruppert, S.; Tollrian, R.; Speck, T. Prey capture analyses in the carnivorous aquatic waterwheel plant (*Aldrovanda vesiculosa* L., Droseraceae). *Sci. Rep.* **2019**, *9*, 18590. [CrossRef]

38. Lebkuecher, J.G.; Eickmeier, W.G. Physiological benefits of stem curling for resurrection plants in the field. *Ecology* **1993**, *74*, 1073–1080. [CrossRef]

39. Rafsanjani, A.; Brulé, V.; Western, T.L.; Pasini, D. Hydro-responsive curling of the resurrection plant *Selaginella lepidophylla*. *Sci. Rep.* **2015**, *5*, 8064. [CrossRef] [PubMed]

40. Ashida, J. Studies on the leaf movement of *Aldrovanda vesiculosa* L. I. Process and mechanism of the movement. *Mem. Coll. Sci. Kyoto Imp. Univ. Ser. B* **1934**, *9*, 141–244.

41. Westermeier, A.; Sachse, R.; Poppinga, S.; Vögele, P.; Adamec, L.; Speck, T.; Bischoff, M. How the carnivorous waterwheel plant (*Aldrovanda vesiculosa*) snaps. *Proc. Roy. Soc. B* **2018**, *285*, 20180012. [CrossRef]

42. Poppinga, S.; Joyeux, M. Different mechanics of snap-trapping in the two closely related carnivorous plants *Dionaea muscipula* and *Aldrovanda vesiculosa*. *Phys. Rev. E* **2011**, *84*, 041928. [CrossRef]

43. Körner, A.; Born, L.; Mader, A.; Sachse, R.; Saffarian, S.; Westermeier, A.S.; Poppinga, S.; Bischoff, M.; Gresser, G.T.; Milwich, M.; et al. Flectofold—A biomimetic compliant shading device for complex free form facades. *Smart Mater. Struct.* **2018**, *27*, 017001. [CrossRef]
44. Liang, H.; Mahadevan, L. Growth, geometry, and mechanics of a blooming lily. *Proc. Natl. Acad. Sci. USA* **2011**, *108*, 5516–5521. [CrossRef] [PubMed]
45. Ridley, H. *The Dispersal of Fruits throughout the World*; L. Reeve & Co., LTD: Ashford, UK, 1930.
46. Armon, S.; Efrati, E.; Kupferman, R.; Sharon, E. Geometry and mechanics in the opening of chiral seed pods. *Science* **2011**, *333*, 1726–1730. [CrossRef]
47. Harrington, M.J.; Razghandi, K.; Ditsch, F.; Guiducci, L.; Rüggeberg, M.; Dunlop, J.W.C.; Fratzl, P.; Neinhuis, C.; Burgert, I. Origami-like unfolding of hydro-actuated ice plant seed capsules. *Nat. Commun.* **2011**, *2*, 337. [CrossRef] [PubMed]
48. Melzer, B.; Steinbrecher, T.; Seidel, R.; Kraft, O.; Schwaiger, R.; Speck, T. The attachment strategy of English ivy: A complex mechanism acting on several hierarchical levels. *J. R. Soc. Interface* **2010**, *7*, 1383–1389. [CrossRef] [PubMed]
49. Abraham, Y.; Tamburu, C.; Klein, E.; Dunlop, J.W.C.; Fratzl, P.; Raviv, U.; Elbaum, R. Tilted cellulose arrangement as a novel mechanism for hygroscopic coiling in the stork's bill awn. *J. R. Soc. Interface* **2012**, *9*, 640–647. [CrossRef]

Article

The Plant-Like Structure of Lance Sea Urchin Spines as Biomimetic Concept Generator for Freeze-Casted Structural Graded Ceramics

Katharina Klang [1,2,*] **and Klaus G. Nickel** [2,*]

[1] Institute of Glass and Ceramics, Friedrich-Alexander-Universität Erlangen-Nürnberg, Martensstraße 5, D-91058 Erlangen, Germany
[2] Department of Geosciences, Applied Mineralogy, Eberhard Karls Universität Tübingen, Wilhelmstraße 56, D-72074 Tübingen, Germany
* Correspondence: katharina.klang@fau.de (K.K.); klaus.nickel@uni-tuebingen.de (K.G.N.)

Abstract: The spine of the lance sea urchin (*Phyllacanthus imperialis*) is an unusual plant-akin hierarchical lightweight construction with several gradation features: a basic core–shell structure is modified in terms of porosities, pore orientation and pore size, forming superstructures. Differing local strength and energy consumption features create a biomimetic potential for the construction of porous ceramics with predetermined breaking points and adaptable behavior in compression overload. We present a new detailed structural and failure analysis of those spines and demonstrate that it is possible to include at least a limited number of those features in an abstracted way in ceramics, manufactured by freeze-casting. This possibility is shown to come from a modified mold design and optimized suspensions.

Keywords: structural graded material; freeze-casting; biomimetic; microcomputed tomography (μCT); porous ceramic; sea urchin spine; alumina; fracture behavior

Citation: Klang, K.; Nickel, K.G. The Plant-Like Structure of Lance Sea Urchin Spines as Biomimetic Concept Generator for Freeze-Casted Structural Graded Ceramics. *Biomimetics* **2021**, *6*, 36. https://doi.org/10.3390/biomimetics6020036

Academic Editors: Thomas Speck and Olga Speck

Received: 1 May 2021
Accepted: 24 May 2021
Published: 31 May 2021

Publisher's Note: MDPI stays neutral with regard to jurisdictional claims in published maps and institutional affiliations.

1. Introduction

Many biological tissues and devices have been systematically studied due to their remarkable engineering properties e.g., the toughness of the coconut endocarp [1,2] or the fire resistance of particular barks [3] with the objective of solving technical problems. In recent years, more and more of these natural materials have been systematically studied with the objective of solving technical problems using abstraction, transfer and application of knowledge gained from biological models [4–6]. Several plants, animals or specific organs of these organisms have evolved strategies to cope with sudden potentially destructive mechanical loads and impacts caused by falling, rock falls, attack of feeding animals and various other environmental loads such as wind gusts or wave movements. The typical mechanism of some biological structures to survive overcritical loads is to dissipate energy by allowing the failure of substructures, whose loss is not endangering the function of the whole organism.

Among the structures to realize simultaneously the properties of structural strength, light weight and fluid transport many plants have developed stems and stalks. Many stems have typical cross-sections [7,8], which show below an epidermis concentric cellular structures with different properties, such as a cortex, a phloem, a xylem and a pith. As an abstracted form this can be viewed as a partially or fully filled tube, hence displaying a core–shell structure with several possible differentiations of the substructures.

While this is commonly found in the plant kingdom, where the cellular construction of stems from organic matter allows many biological functions, we do not associate these structures with brittle natural materials. However, the structure of sea urchin spines, although being part of an animal, is remarkably close to plant stems. In this paper we focus on the aboral spines of *Phyllacanthus imperialis*. A cross section from such a spine is shown in Figure 1 to illustrate the plant-alike structure on a micrometer level. Three different cellular concentric structures are characteristic for the spine in total differentiating in their arrangement and density within the microstructure. The building material of the three-dimensional skeletal meshwork is high-magnesium calcite ($Ca_xMg_{1-x}CO_3$) and is known as stereom [9,10]. Investigations have shown that the $MgCO_3$ content varies between 2 to 12 mole percent in the stereom structures of the spine. The stereom structure in the center of the spine is the medulla, which is characterized by a highly permeable cellular structure. Denser cellular stereom structures enclose the medulla being defined as radiating layer. The radiating layer is encircled by an almost dense shell—the cortex. The pores of the stereom are filled with stroma consisting of fluids and several organic compounds such as extracellular fibrils and connective tissue cells [11].

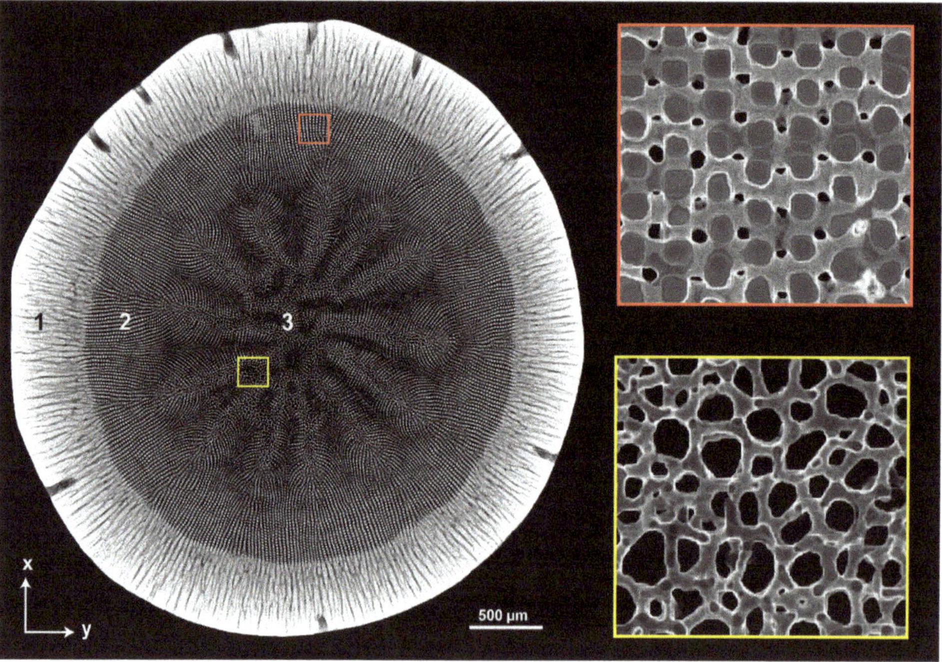

Figure 1. Cross sectional image of the aboral spine of *Phyllacanthus imperialis* based on µCT data visualizing the core–shell structure with its differentiated substructures. 1: Cortex, 2: Radiating layer, 3: Medulla.

On the nanometer, scale the stereom meshwork consists of highly ordered polycrystalline mesocrystals [12]. These mesocrystals are characterized by highly ordered nanoparticle-sized domains, oriented parallel to the z-axis. Small quantities of organic macromolecules (<1 wt.%) are within the magnesium calcite domains [13]. The hierarchical organized lightweight construction of the spine is characterized by remarkable mechanical properties such as high strength [14,15] and beneficial failure behavior dissipating energy in large quantities [16] despite their cellular, brittle microstructure. The transfer of such desirable properties such as strength, high energy dissipation capacity and lightweight into ceramic-based materials and structures is of particular interest for the area of separation technology/filtration (e.g., particle filtration, liquid filtration), in chemical and thermal process engineering (e.g., catalyst carriers, pore burners), medical technology (e.g., bone substitute) and in lightweight construction under high pressure.

It has been shown in [17] that the manufacturing of a biomimetic porous ceramic containing cap-shaped layers of alternating porosities is basically feasible utilizing starch-blended slip casting. This modified slip casting method has its limitations. Ceramics manufactured with this method lack the homogenous and ordered pore geometry. Other techniques like the layer-by-layer deposition [18,19], filtration [20,21], hydrogel casting [22] and electrophoretic deposition [23] are time-consuming and size limiting processes, which enable to fabricate ceramic layers with a thickness of approximately 200 µm only. Freeze-casting produce, in contrast to the techniques mentioned above, porous ceramics with complex, three-dimensional pore structures that can also contain at least two different levels of open porosity and an ordered pore alignment. Mimicking those features is required in order to manufacture the abstracted concepts of the spine of *Phyllacanthus imperialis* into a porous bioinspired ceramic. In general, the freeze-casting technique has clearly become a focus of attention in the last decade concerning the manufacturing of complex pore geometries and open interconnected pore systems due its simple operation, cost efficiency and environmental friendliness.

The manufacturing process uses a ceramic suspension, which is directional frozen and then sublimated before sintering (Figure 2). This process provides materials with a unique porous architecture, where the porosity is almost a direct replica of the frozen ice crystals.

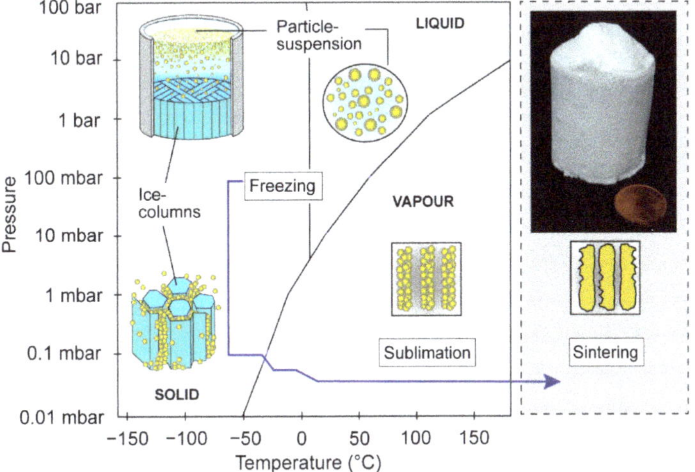

Figure 2. Schematic overview of the freeze-casting process with four main processing steps: preparation of the suspension, solidification via freezing, sublimation and sintering.

The pore morphology, orientation and average pore size can be tailored by altering process parameters such as slurry concentrations [24–26], freezing temperatures [25,27–29], cooling rates [25,30–32] and use of additives [25,26,33]. It is important that the prepared slurries are stable during the entire duration of the freezing stage. Since the solvent initially present in the slurry is converted into solid, that is later eliminated to form the porosity in the ceramic, the pore content can be adjusted by tuning the slurry concentrations. The porosity of the ceramic is directly related to the volume of the solvent. A wide range of porosities, approximately from 25 to 90%, can be achieved via freeze-casting [25]. The total porosity is also depended on numerous additional parameters affecting the packing of particles between the solvent crystals such as the nature of the solvent, its viscosity, the particle morphologies and size distribution [26]. The solidification behavior of the freezing vehicles and thus the pore structure left by the frozen vehicles is affected by the freezing temperatures [25,27–29]. The pore channel size decreases significantly with lower freezing temperatures, regardless of any possible microstructure variations in the individual specimens. Porosity also decreases as freezing temperatures decline. With decreasing freezing temperatures, the solidification velocity is increased inducing smaller lamellar ice crystal spacing and thus pore channel size. This freedom adds significant difficulties when trying to understand the underlying principles that govern the relations between processing and the obtained microstructure. The pore structure of freeze-cast ceramics is determined by the morphology of the solidified fluid [34]. Water is the most common fluid for the freeze-casting process, because of its environmental friendliness, low cost, easy handling and best application chances. Furthermore, employing a unidirectional (=conventional) freeze-casting technique, lamellar pore structures can be fabricated. Generation of wall interconnectivity during the solidification can be attained by modifying the character of the suspension. This can be achieved in a water-based suspension by using additives. To suppress the preferential lamellar growth of ice crystals extensively, it is known from previous studies [24,35] that gelatin as additive in the conventional freeze-casting procedure induces a cellular pore morphology exhibiting an increased interconnectivity of the cell walls. The water molecules, surrounding the gelatin, form a discontinuous network and prevent the concurrent preferential lamellar growth of ice crystals. Instead, spatially separated ice nuclei are formed in the gelatin network. Thus, the ice crystals are reduced in size and show a polygonal morphology [24].

Using the principle of the bottom-up approach of biomimetics [36], we analyze in-depth the lance sea urchin spine *Phyllancanthus imperialis* quantitatively in terms of its form-microstructure-function relationship and its mechanical properties to understand the operating principles of the relevant structures. These are interesting for various technical implementations, i.e., structurally graded ceramics with anisotropic pore morphologies being used as particulate filters for vehicular emission control. High-resolution micro-computed tomography (μCT) proved to be ideally suited to analyze the internal structures of the spine non-invasively regarding the strut configuration, pore sizes and porosities. Based on the results of the relation between the mechanical and microstructural properties, we present a new detailed structural analysis and failure observation of those spines and demonstrate the inclusion of a limited number of those abstracted features in freeze-casted ceramics. A modified mold design and optimized suspension was used in the freezing process to obtain the gradation features in the ceramics.

2. Experimental Procedure

2.1. Lance Sea Urchin Spines

The sea urchins were not killed for the investigation of the spine material. The investigated primary spine material belongs to the tropical cidaroid *Phyllacanthus imperialis* [37]. The spine material was purchased from a material supplier (Mineralien- und Fossilien-handlung Peter Gensel, Weimar, Germany).

2.2. Freeze-Casting

Unidirectional alumina scaffolds inspired by the spine's microstructure were manu-factured via the freeze-casting route from a water-based alumina suspension. Gelatin as additive was used in the water-based suspension. Water suspensions of gelatin powder (Ewald-Gelatin GmbH, Bad Sobernheim, Germany) were prepared with a solid loading of 6.8 and 3.5 vol.% (Table 1). Homogenization of the gelatin suspension was carried out at 60–70 °C by stirring for 1 h. Under continuous stirring of the gelatin suspension, the 99.99 % α-alumina powder (TM-DAR Taimicron of Taimei Chemicals Co, d50 = 1.2 μm, $\rho = 3.98$ g/cm^3) was added slowly in small measures to obtain a solid load of 16.5 vol.% by dispersing them with 0.47 cm^3 of Dolapix CE 64 (Zschimmer & Schwarz GmbH & Co KG Chemische Fabriken, Lahnstein, Germany), respectively (Table 1). The volume of Dolapix was fixed at 0.5% of the solid content (recommendation of the manufacturer). The dispersion was carried out with an ultrasonic processor UP400S (Hielscher Ultrasonics GmbH, Teltow), equipped with 20 mm titanium sonotrode (Amplitude 50%, cycle 0.5 s), for every alumina–gelatin suspension for 5 min. A constant rotation of the beaker at the titanium sonotrode ensured a homogeneous distribution of the alumina particles and temperature. The temperature of the alumina–gelatin suspension was monitored and kept below 80 °C to avoid hydrolysis of the gelatin.

Table 1. Concentration of the solid loading of Al_2O_3 and gelatin of the water-based alumina suspensions.

Name	φ Sintered Foam	Total Concentration of Al_2O_3		Total Concentration of Gelatin with Regard to Water		Dolapix CE 64
	(vol.%)	(wt.%)	(vol.%)	(wt.%)	(vol.%)	(cm^3)
Polygonal cells	77–79	44.0	16.5	5	6.8	0.47
Oblate cells	77–79	44.0	16.5	2.5	3.5	0.47
Graded ceramic	77–79	44.0	16.5	5	6.8	0.47

Polytetrafluorethylene (PTFE) tubes (with an inner diameter of 25 mm and a height of 35 mm) were used as molds that were placed on a copper plate with a thickness of 3 mm (Figure 3A). Before filling the alumina suspension into the PTFE tubes, the copper plate was cooled at a temperature of −60 °C. After 4 h of directional freezing, the frozen specimen was carefully removed from the PTFE tubes by slightly heating the PTFE tubes (Figure 3A). The water-based ice crystals were removed via sublimation. The obtained green bodies were sintered in a conventional furnace. Each intermediate temperature was achieved with a constant ramp speed of 1 °C/min. In the first stage the temperature was increased to 550 °C and dwelled for 2 h; subsequently the specimens were heated to 900 °C and dwelled for 2 h. Finally, the target sintering temperature of 1350 °C was reached and held for 4 h.

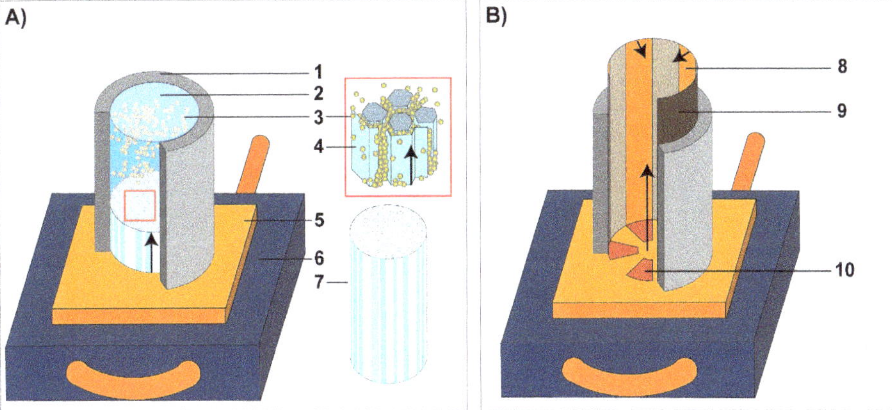

Figure 3. Schematic illustration of the mold designs for the freeze-casting process. (**A**) The conventional setup for producing unidirectional pore systems consists of a polytetrafluorethylene (PTFE) tube that was fixed on a copper plate. The mold design in (**B**) was used to manufacture a structural graded ceramic inspired by the microstructure of spine of *Phyllacanthus imperialis*. The freezing direction is indicated by the black arrows. 1: PTFE-mold, 2: particle suspension, 3: Al_2O_3 particle, 4: ice crystal, 5: copper plate, 6: cold plate, 7: frozen suspension with unidirectional ice crystals, 8: copper film, 9: Teflon film, 10: silicon film.

The basic mold design was manipulated through the addition of silicon and copper plates (Figure 3B). To test the manipulated mold design, the slurry composition corresponding to the ceramic 'Polygonal cells' (see Table 1) has been used for this purpose. The manipulated mold design consists of triangle-shaped silicon slices being arranged in a circular array and alternating with the copper plate at the bottom of the mold (Figure 3B). Alternating copper stripes (thickness = 500 μm) are placed at the side walls.

2.3. Sample Preparation and Characterization

The spine segments were characterized using optical light microscopy and scanning electron microscopy (SEM). For determining the quantity of the medulla, radiating layer and cortex in the spine segments, only the digital light microscope (Hirox RH-2000, Hirox Europe Ltd., Limonest, France) was used. The dimensions (length, diameter) of the spine segments were determined with a digital caliper. The microstructure of uniaxial loaded core cylinder of the spine, however, was investigated utilizing the SEM. After the first significant acoustic signal in the initial stage of loading, the uniaxial compression was stopped to examine the cracks in the microstructure of the core cylinder. The porosity of the spine segments was determined gravimetrically by assuming a density, ρ, of 2.711 g/cm^3 for calcite [38]. Therefore, the volume of the samples was calculated by measuring the area of the cross section, A, on the microscopic images and multiplying it with the mean height, h. Together with the mass, m, the porosity was estimated from the following equation:

$$\phi(\%) = \left(1 - \left(\frac{m(\text{g})}{A(\text{cm}^2)\cdot h(\text{cm})\cdot\rho(\text{g}\cdot\text{cm}^{-3})}\right)\right) \times 100 \tag{1}$$

The microstructure of the fabricated ceramics was investigated using SEM. Pore sizes (pore axis lengths, pore diameters) were determined utilizing the software Fiji 2.0.0 (ImageJ). In total, 100 pores were considered for each ceramic species. The porosity of the ceramics was determined using Equation1). Instead of the density of calcite, a theoretical density of α-Al$_2$O$_3$ of 3.97 g/cm^3 [39,40] was implemented in Equation (1).

2.4. Scanning Electron Microscopy (SEM)

Spine segments (cut parallel and perpendicular to the z-axis) and ceramic samples were investigated with a scanning electron microscope (Tabletop Microscope TM3030plus, Hitachi Ltd., Tokyo, Japan) at the Institute of Geosciences of the University of Tübingen. The backscatter electron mode (BSE) has been applied for the microscopic analysis at an acceleration voltage of 15 kV and at a working distance of 6500 µm.

2.5. Micro Computed Tomography (µCT) and Data Processing

The upper part of a spine (Figure 4, red box) was scanned in a nano CT scanner (phoenix nanotom m, Fa. GE Sensing & Inspection Technologies GmbH, Germany) at the Institute of Textile Technology and Process Engineering Denkendorf (ITV). The samples were scanned at 800 kV/180 µA with an exposure time of 800 ms to a resolution of 1.27 µm per voxel. The acquired two-dimensional X-ray images were reconstructed with a Filtered Back Projection reconstruction algorithm. The µCT data are analyzed and reconstructed into three-dimensional objects using the Fiji 2.0.0 (ImageJ) software package. In order to determine the porosity xy-images were binarized using the iterative 'default' threshold based on [41]. Based on this binarization process, the software calculates the amount of black and white in the image. The algorithm of the plugin BoneJ [42] has been used to determine the thickness of the struts. For this purpose, binarized images have been utilized for the evaluation of the thickness as well. The pore axis lengths were determined from the reconstructed three-dimensional objects applying the mode 'fit ellipse'.

2.6. Mechanical Testing

Uniaxial compression tests were conducted in a universal testing machine (Instron 3180, Instron Deutschland GmbH, Pfungsstadt, Germany). The spine segments and ceramic cylinders were placed on a Si$_3$N$_4$ plate and pressed against a tungsten carbide compression die. The force was measured simultaneously by a force transducer. A photograph and schematic view of the measuring principles of the universal testing machine is illustrated in [16]. All experiments were performed with a crosshead movement speed of 0.5 mm/min. During the uniaxial compression tests of the specimen, the fracture behavior and the displacement was monitored by a video extensometer (LIMESS RTSS-C02, LIMESS Meßtechnik GmbH, Krefeld, Germany).

From the geometry of the spine specimens, the maximum compressive strength has been calculated from the load at the first crack formation. The displacement recorded from the video extensometer was used for calculating the Young's modulus from the slope of the linear elastic increase in the stress–strain plots.

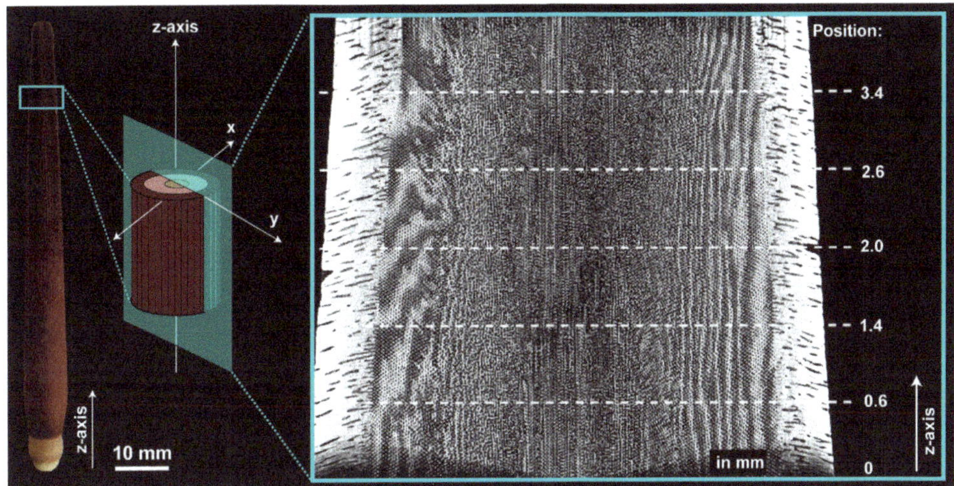

Figure 4. Spine of *Phyllacanthus imperialis*. Turquoise box illustrates the scanned area via µCT. A complete reconstruction of the scanned spine area can be seen on the right-hand side.

3. Results

3.1. Sea Urchin Spine as Concept Generator

The microstructure of the adult, aboral spine of *Phyllancanthus imperialis* is composed of three different concentric structures being referred as medulla, radiating layer and cortex from the inside to the outside (Figure 5A). The cortex is a dense, but permeable shell that differentiate essentially from the cellular stereom arrangement of the medulla and radiating layer. A stacked µCT section of the cortex reveals several pore channels within the microstructure being oriented perpendicular to the z-axis (Figure 5B). Besides the small pore channels, the external morphology of the cortex contains also comparably large pores appearing very irregularly in their shape. Their pore channels, however, cannot be pursuit deeply into the internal microstructure of the cortex and disappear in the outer half of the cortex (Figure 5B). The quantity of the pores in the microstructure of the cortex increases visually from the outside to the inside, which is associated with an increase of the porosity from the outside to the innermost cortex structure: from an average of $11 \pm 0.9\%$, the porosity increases up to an average of $18 \pm 1.0\%$ (Figure 6A). In total the cortex is characterized by an average porosity of $15.2 \pm 3.2\%$ (Figure 6A). The external surface of the cortex is characterized by an average pore size of 28 ± 10 µm (Figure 6B). Smaller pore sizes can be found in interior structures of the cortex: the average pore size is constant within the interior and varies between 15 ± 8 µm and 17 ± 8 µm (Figure 6B).

The µCT reconstructions and sections visualize the inner mesh structure of the radiating layer and medulla (Figure 7). The radiating layer composes of thin sheets of stereom layers being separated by a regular arrangement of cross struts. The separation distance between the sheets appears regular and approximately equal to the thickness of the individual sheets. Each individual sheet is permeated by oval-shaped channels appearing in their multitude very even. Each sheet comprises a strict arrangement of pore rows parallel to the z-axis. The rows of pores are, however, arranged with a minimal misalignment along the z-axis meaning that every second pore row is congruent. The cross struts connecting each individual stereom layer are orientated perpendicular to the stereom sheets and are arranged in a periodic way. Every second row of cross struts is congruent analogously to the arrangement of pores. An analysis of the pore axis lengths (Figure 8A), a_1 and a_2, of the pores on the zx- and zx-plane (Figure 7) indicates similar average values (zx-plane: a_1 = 17.7 µm, a_2 = 10.1 µm; zy-plane: a_1 = 16.8 µm, a_2 = 9.5 µm).

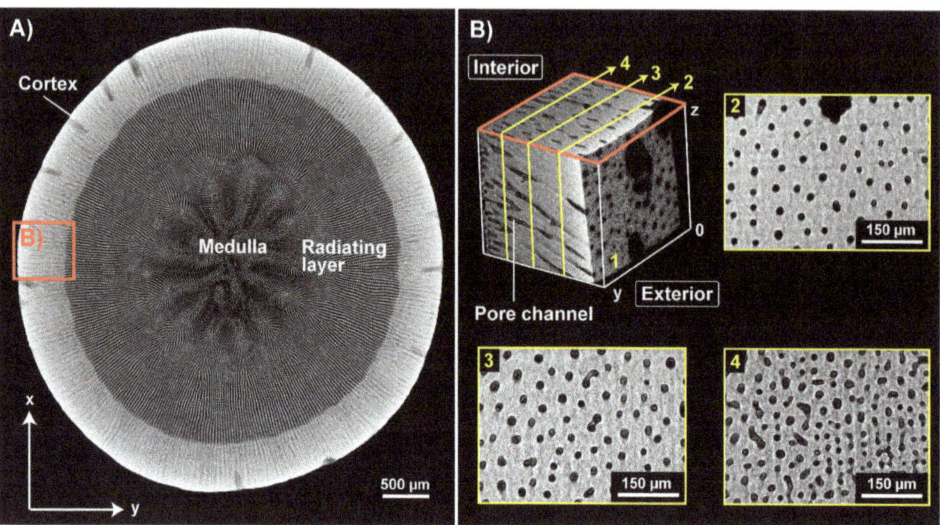

Figure 5. Microstructure of the spine of *Phyllacanthus imperialis*. (**A**) Z-projection of the microstructure of the spine shows the three structural units: the cortex, radiating layer and medulla. Various sections (yellow boxes) of the cortex microstructure are displayed in (**B**).

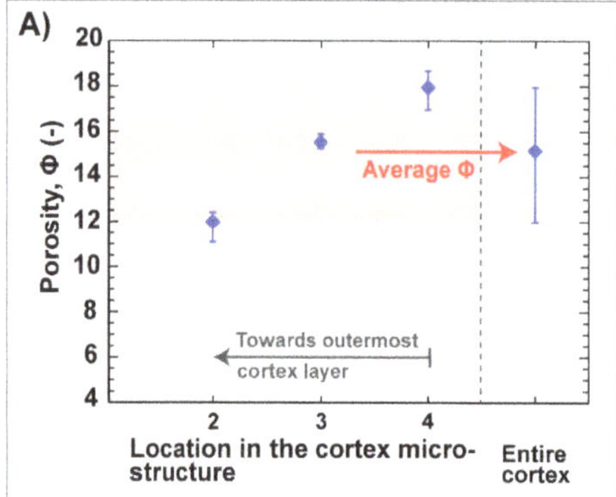

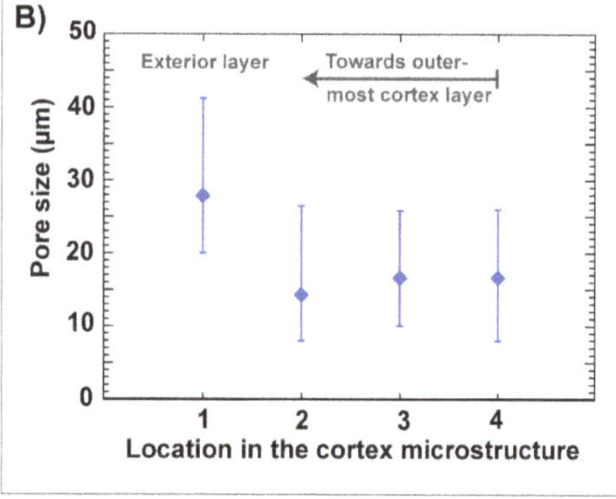

Figure 6. Porosity, Φ, of the cortex and the size of the pore channels within the cortex of the spine of *Phyllacanthus imperialis*. The porosity of the µCT sections of the cortex is displayed in (**A**). The average porosity of each µCT section is displayed in (**B**). The assignment of the µCT sections (2, 3, 4) is displayed in Figure 5B.

The spine microstructure is characterized by a columnar arrangement of the stereom mesh (Figure 7). The stereom mesh consists of several long trabecular rods, which are linked by cross struts in such way to produce irregular polygonal passages parallel to the z-axis. The number of trabecular rods forming an individual columnar stereom varies from 4–8. The lateral pores on the zy-section are regularly formed and arranged in an offset manner to the adjacent pores. The average pore axis lengths of the polygonal passages (a_1 = 22.0 µm, a_2 = 17.4 µm) on the xy-section and of the pore rows on the zy-axis (a_1 = 19.6 µm, a_2 = 14.7 µm) are larger compared to the pores from the layered stereom type

(Figure 8A). The smaller difference between the average pore axis lengths of the columnar rows indicates that the pores have a rather circular than an oval character.

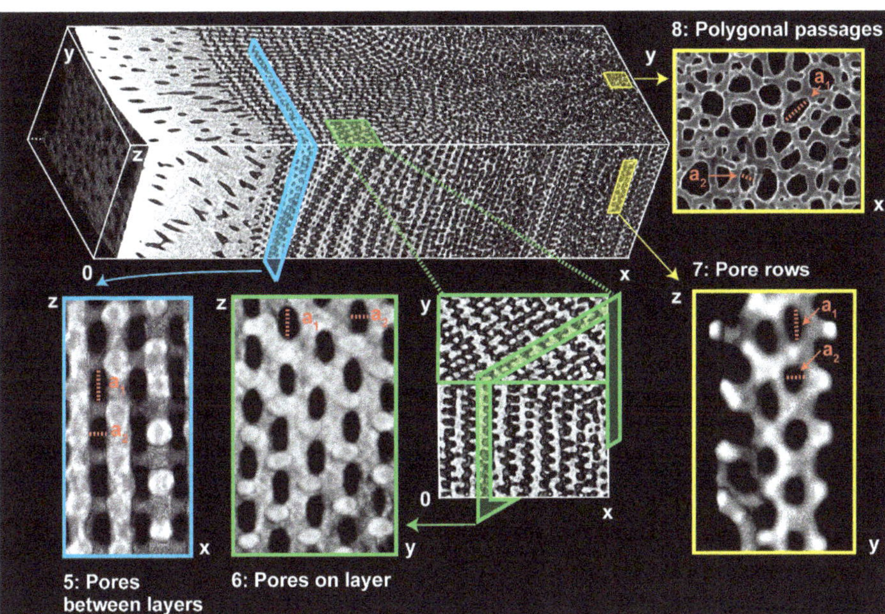

Figure 7. A stacked µCT reconstruction of the inner microstructure of the spine of *Phyllacanthus imperialis*. The turquoise and green boxes visualize different cross-sectional planes of the structure of the radiating layer. The inner mesh of the center of the spine is shown in the yellow boxes (polygonal passages, pore rows). The largest and the smallest pore diameter are indicated by a_1 and a_2, respectively.

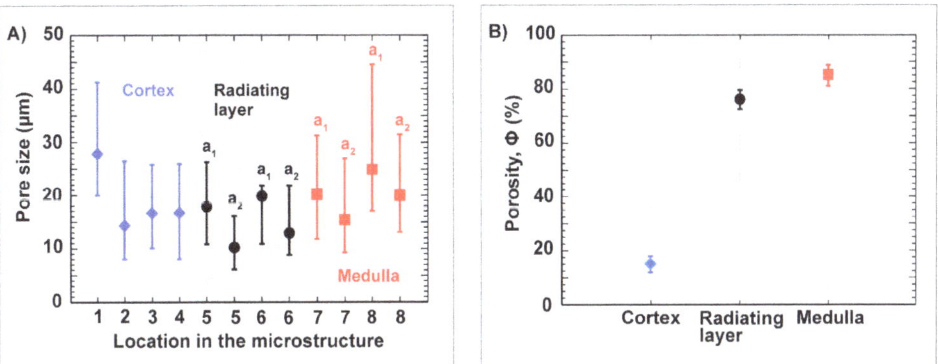

Figure 8. Characterization of the pore sizes and porosities of the microstructural units (cortex, radiating layer and medulla) in the spine of *Phyllacanthus imperialis*. The pore axis length is shown by a_1 and a_2, and of the structures forming the radiating layer, the medulla and the cortex are displayed in (**A**). The location within the spine microstructure is given as numbers. The affiliation of the numbers is shown in Figures 5B and 7, respectively. The plot in (**B**) represents the average porosities of the cortex, radiating layer and medulla.

The average porosities of the three structural units of the spine are demonstrated in Figure 8B. The stereom mesh in the center of the spine, the medulla, is characterized by an average porosity of $85.7 \pm 2.8\%$. The dense, regular arrangement of the stereom sheets in the radiating layer has an average porosity of 76.6 ± 3.0%. In comparison, the cortex is the densest of the three structural units and was characterized by a porosity of 15.2 ± 3.2%.

The calculated strut thickness distribution of the microstructure indicates that the stereom sheets from the radiating layer appearing as lateral strut routes (on the xy-plane) and the polygonal passages from the medulla are clearly separated from each other (Figure 9A). Thus, there is no transition zone regarding the strut thickness between both structural types. There is a clear border between the polygonal passages and the lateral strut routes: the polygonal passages are characterized by a strut thickness of <10 μm, whereas the stereom sheets have a thickness of >10 μm (Figure 9A,B). Based on the μCT data, the average strut thickness for the stereom structures in the medulla and radiating layer has been determined to be 8 ± 1.6 μm and 13.7 ± 2.0 μm, respectively.

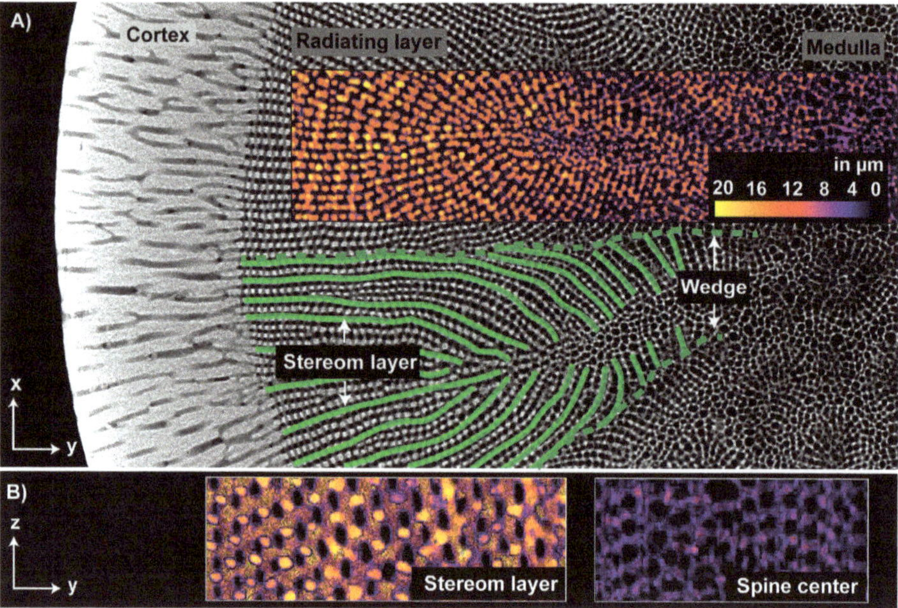

Figure 9. (**A**) μCT reconstruction of the spine of *Phyllacanthus imperialis* demonstrating the calculated strut thickness distribution of the meshes (stereom layers and polygonal meshes in the spine center) and a superstructure. The stereom layers (marked in green) running directly from the cortex to one sidearm of the spine center (xy-plane). Thus, a wedge-like structure (dashed line in green) is formed, which is a superstructure. (**B**) Cross sections, parallel to the z-axis, of the stereom layer and spine center. The color scheme in (**A**,**B**) indicates the strut thickness.

The mutual distribution of the two stereom structures in spine microstructure forms a superordinate structure, which is here described as a superstructure (Figure 9A). In this case, the strut configuration is slightly modified resulting in a subordinate structure, which is not obviously visible on every cross-sectional level. This kind of superstructure can be limited to a particular cross-sectional level and is therefore visible on this level. Taking the stereom structures of the radiating layer and medulla together, a superstructure becomes visible on the xy-plane (Figure 9A). Each individual stereom sheet is linked with the cortex microstructure. Starting from the cortex, each stereom sheet runs directly to the nearest side branch of the medulla in a straight to slightly curved manner. Consequently, a group of several stereom sheets forms a 'wedge-like' structure that encloses a side branch of the medulla. This sort of subordinate structure, where a group of stereom sheets encloses a side branch of the medulla, and is here called a 'wedge' (Figure 9A). In total, 13 wedges have been determined in the adult, aboral spine of *Phyllacanthus imperialis*.

To quantify the structural relationship of the spine segments with the mechanical behavior in uniaxial compression, the spine microstructure was divided into a shell (=cortex) and compliant core (=radiating layer and medulla). A simplification of the spine microstructure into a core–shell structure helps to model the influence and behavior of the cortex in the spine segment during uniaxial compression. A 'structural factor' was, therefore, created to summarize the quantity of the cortex in the spine segments (Figure 10A). The structural factor is composed of the radius of the spine segment, a, which was normalized by the cortex thickness, t (Figure 10A: marked in red). Such a structural parameter, expressed as radius to thickness ratio a/t, was first established and utilized by [43–45], who used the a/t ratio to characterize and model the shell-core system of plant stems, animal quills and bird feather rachis in terms of their mechanical efficiency. The a/t ratios of the structural elements in cross section (xy-plane) over the entire spine length is summarized in Figure 10B (= black squares). The a/t ratio increases from the spine tip to the base. Low a/t ratios are, in principle, an indication for a large cortex thickness, which is clearly the case for the spine tip. A large proportion of the cortex is, therefore, associated with low values of the porosity in the spine segment (Figure 10B, red circles). With a decrease of the cortex thickness, the porosity increases constantly up to a value of approximately 67% (Figure 10B, red circles).

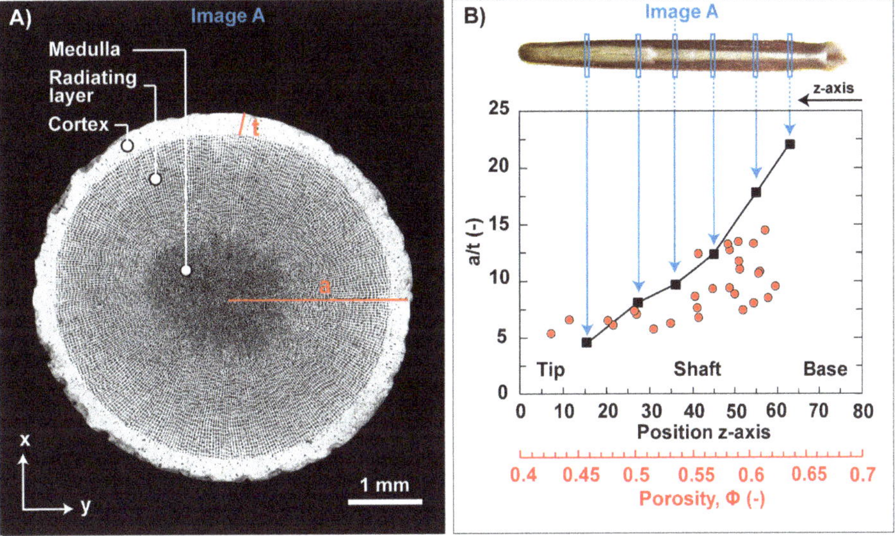

Figure 10. Quantity of the cortex in the microstructure of the spine of *Phyllacanthus imperialis*. (**A**) The quantity of the cortex is expressed as a/t ratio. The a/t ratio is composed of the radius of the spine segment, a, which was normalized by the cortex thickness, t. The quantity of the a/t ratio and the porosity at specific positions in the spine is demonstrated in (**B**) as black squares and red circles, respectively.

The dependency of the Young's modulus and maximum compressive strength on the microstructural core–shell design and accompanied porosities of the spine segment is displayed in Figure 11. The color bar indicates the a/t ratio of the measured, uniaxial loaded spine segments. The maximum compressive strength (Figure 11A) and Young's modulus (Figure 11B) is negatively correlated with the porosity of the spine segments. It means in relation to the core–shell design that a decrease of the a/t ratio in the microstructure is associated with an increase of the Young's modulus and maximum compressive strength. A decrease of the a/t ratio is accompanied with a rise of the volume fraction of the cortex. The volume fraction of the cortex influences also the fracture behavior of the spine segments. Two main failure modes have been identified, named failure mode I and II (Figure 11).

Spine segments undergo failure mode I are characterized by a/t ratios of smaller than 9.2. Failure mode II occurs in spine segments being marked by large a/t ratios of more than 9.2. The change of the failure modes is in the porosity range of approximately 55% (Figure 11: light-yellow line).

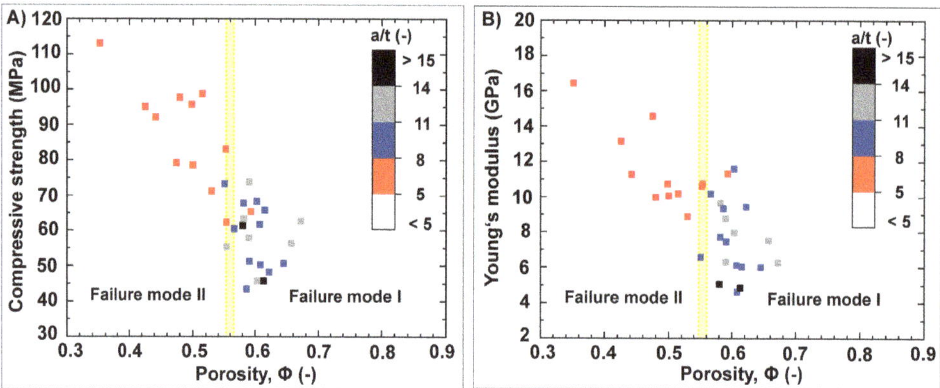

Figure 11. Plots in (**A,B**) show the maximum compressive strength and Young's modulus of the spine segments of *Phyllacanthus imperialis* as a function of the porosity, Φ. The color bar indicates the structural factor of the spine segments, the a/t ratio. A change of the failure modes was observed at a porosity of approximately 55% being highlighted in the plots as light-yellow line.

Typical stress-strain curves and corresponding microphotographs of the failure modes of the spine segments are shown in Figure 12. Failure mode I shows a high plateau strength after the maximum compressive strength, which is comparable to the failure behavior of cellular materials in general (Figure 12A). A sharp decrease of the stress after the maximum compressive strength followed by a constant low stress level are characteristic features of failure mode II (Figure 12B). The core–shell system reacts to the compressive stress, which dependents on the volume fraction of the cortex in spine segment. Failure mode I is characterized by a rigid connection of the cortex-core system, because the cortex remains attached to the core during the entire compressive loading. A breakdown of the core–shell system can be observed in failure mode II, since the cortex flakes off the spine structure after reaching the maximum compressive strength of the spine segment.

In detail the fracture behavior of failure mode I are as follows: after the subsequent linear elastic increase, the first drop in stress is accompanied by the generation of individual horizontal cracks, which occur in the lower region of the spine segment (Figure 12A,b). Originating from the horizontal ring crack, vertical cracks initiate to propagate through the spine segment (Figure 12A,c). The stress oscillates on a plateau level after crushing of the lower region between the bottom and ring crack by continued spallation. Simultaneously, a lath-like segment splits off from the spine structure, which is relatively small compared to the remaining spine segment (Figure 12A,d). A significant drop in the stress can be observed after complete spallation of the lath-like segment (Figure 12A,e). After more than 50% crushing of the remaining spine segment, a significant increase in the stress is observable and was characterized as densification phase (Figure 12A,f).

Failure mode II is characterized by an additional, brief period in which the cortex includes its own failure behavior before reaching the maximum compressive strength of the spine segment (Figure 12B). The cortex starts to dismantle slowly from the spine structure after the linear elastic increase, which can be observed, since the coloring of the cortex appears to be lighter in the microphotographs compared to the undamaged cortex (Figure 12C,b). The dismantling of the cortex occurs at 65 MPa being an inflexion point in terms of the stress gradient: the gradient of the stress curve levels off after the dismantling of the cortex. The dismantling balanced the maximized stress acting on the core-cortex interface (Figure 12C,c). The longitudinal spallation of the cortex in large quantities occurs after the maximum compressive strength has been reached. Thus, the core becomes exposed (Figure 12C,d). Horizontal and vertical cracks start to propagate simultaneously through the core within few milliseconds (Figure 12C,e). A significant decrease in the stress can be observed due to the complete crushing of the upper section of the core. Several lath-like segments separate thereafter from the remaining core (Figure 12B,f). A vertical tilting of the lath-like segments occurs with continuing compression; therefore, the stress drops again. The last stage of failure mode II is also characterized by a densification of the remaining segments (Figure 12B,g).

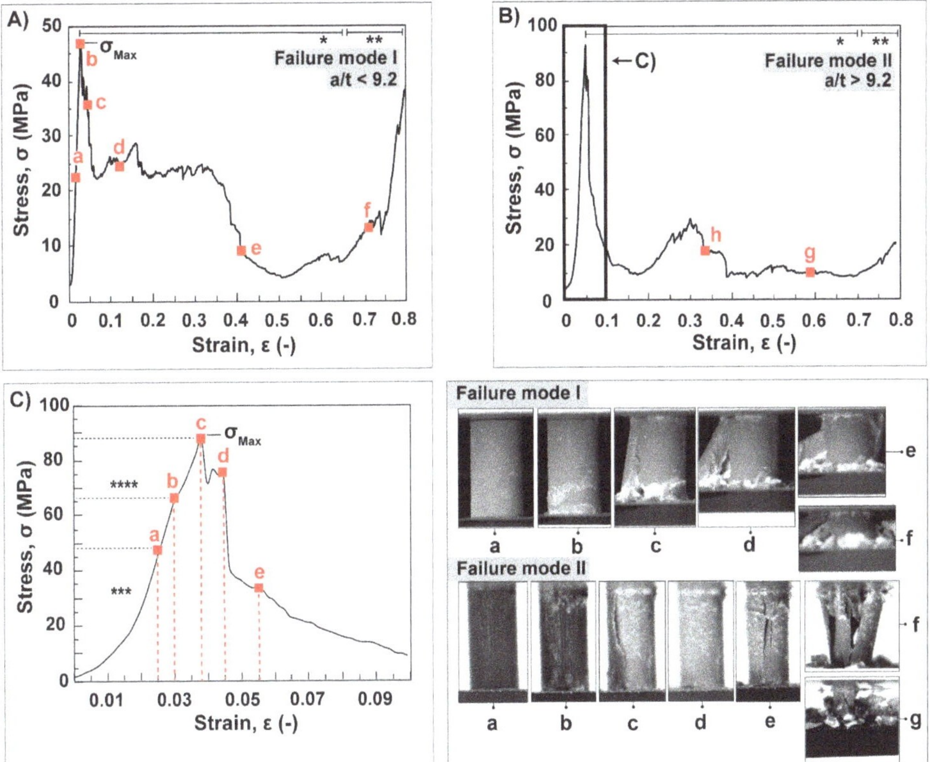

Figure 12. Spine segments of *Phyllacanthus imperialis* under uniaxial compression. Typical stress (σ)-strain (ε) curves of failure mode I and II are demonstrated in (**A**,**B**), respectively. A detailed view of the initial stress–strain curve of a spine segment undergoing failure mode II is displayed in (**C**). The red letters belong to the microphotographs presenting the fracture behavior at a specific stage of uniaxial compression. *: 'Quasi-ductile' regime. **: Densification. ***: Linear elastic regime with a brief alignment period. ****: Dismantling of the cortex.

The specific energy absorption per volume unit has to be considered in order to quantify the mechanical effectiveness of the failure modes from the spine segments. The parameter $U_V(\varepsilon)$ is composed of the ratio of the energy absorption, $W(\varepsilon)$ and volume of the spine segment, $V(\varepsilon)$, destroyed during compression:

$$U_V(\varepsilon) = \frac{W(\varepsilon)}{V(\varepsilon)} \tag{2}$$

The energy absorption, $W(\varepsilon)$, is a measure of the total amount of energy that was dissipated during straining and calculated by integrating the area under the stress-strain curve, which is given as Equation (3):

$$W(\varepsilon) = \int_0^\varepsilon \sigma(\varepsilon)\,d\varepsilon \tag{3}$$

A normalized value of the energy absorption capacity of materials is the energy absorption efficiency, which was adapted by metal foams literature. The energy absorption efficiency, $\eta_{(\varepsilon)}$, characterizes a materials ability to absorb stress as it compresses by normalizing the energy absorbed by the maximum stress observed [46]:

$$\eta_{(\varepsilon)} = \frac{1}{\sigma_{max}\varepsilon} \int_0^\varepsilon \sigma\,d\varepsilon \tag{4}$$

The peak value of the stress during uniaxial compression is designated as σ_{max}. Figure 13A shows the average values of the energy absorption per volume unit at the strain of 0.3, 0.5 and 0.7 calculated by Equation (2). Their normalized average values according to Equation (4), expressed as energy absorption efficiency, are displayed in Figure 13B. The average values of the energy absorption efficiency of failure mode I drops to an average value of 65% at a strain of 0.7 (Figure 13B, black squares). A huge decrease of the average values of the energy absorption efficiency can be determined for the spine segments undergoing failure mode II: the average values of the energy absorption efficiency decrease down to 30% at a strain of 0.7 (Figure 13B, red squares).

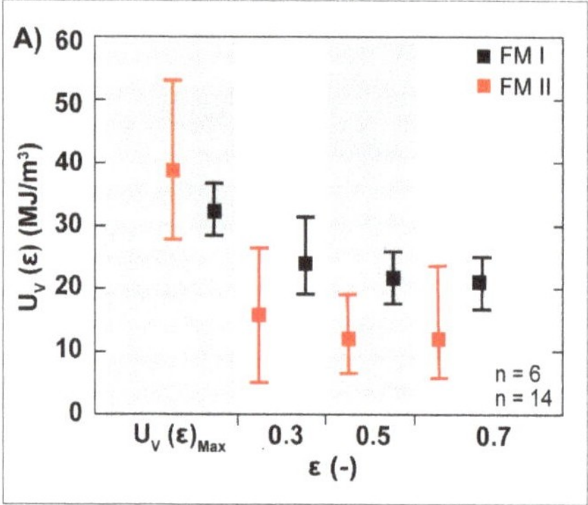
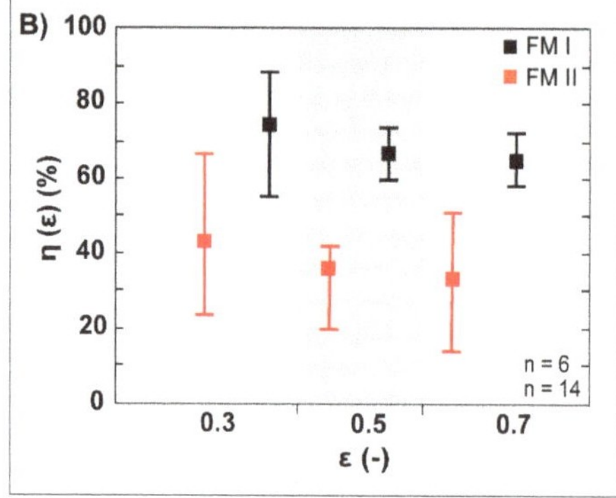

Figure 13. Specific energy absorption per volume unit, $U_V(\varepsilon)$, and the energy efficiency, $\eta(\varepsilon)$, of the spine segments of *Phyllacanthus imperialis*. The relation of the average values of the $U_V(\varepsilon)$ and $\eta(\varepsilon)$ to the strain, ε, is displayed in (**A,B**), respectively. FM I: Failure mode I, FM II: Failure mode II.

Both failure modes are characterized by a quasi-ductility based on multiple fracturing indicating that the spine segments are able to dissipate energy through the generation of a horizontal and vertical crack system and segmentation. Both failure modes differentiate in terms of the intensity of the structural segmentation of the spine segments. During compressive loading, the exposed core (failure mode II) is divided structurally into several lath-like segments. Therefore, the structural integrity is no longer ensured and the capability to absorb energy is, thus, reduced greatly. In contrast to that, the rigid core–cortex connection (failure mode I) is able to reduce the intensity of the structural segmentation. A complete segmentation of the spine segment does not occur in this case. The major proportion of the spine segment remains the same, even though small lath-like segments flake off laterally. Since the quantity of the remaining material is larger compared to the spine segments characterize by failure mode II, the spine segments are, therefore, still able to dissipate energy even at high strain levels. This, in turn, results in comparable large values of the energy absorption per volume unit and energy absorption efficiency for spine segments being subjected to failure mode I. The cortex is able to keep the structural integrity of the spine segments even at advanced stages of compression.

The failure modes are based on the microstructural configuration of the spine segments. Both failure modes include the mode of segmentation as failure mechanism. Uniaxial experiments of prepared core cylinders (= medulla, radiating layer) reveal the relationship between the strut configuration and the propagation of cracks. For observing the crack propagation in the core cylinder, the uniaxial loading was stopped after the first acoustic signal occurred. Compression experiments on the core cylinder disclose two stages of crack propagation: initial and advanced stage of crack propagation (Figure 14). The first cracks start to propagate directly along the border between the medulla and radiating layer (Figure 14, red box), which is clearly marked as initial stage of the crack propagation. It indicates that the stress is mainly concentrated at the interface between the medulla and radiating layer. Continuing the compressive loading on the core cylinder, the cracks propagate outwards along the stereom sheet of the radiating layer (Figure 14). Therefore, a segment spalls off from the structure and the inner spine microstructure remains unchanged. No cracks appear in the spine's inner microstructure.

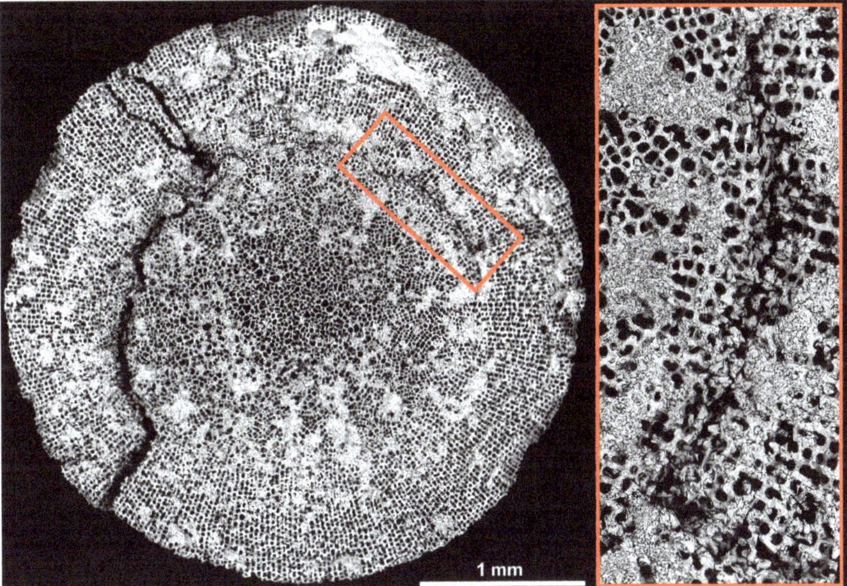

Figure 14. Cross sectional BSE image of the spine's core of *Phyllacanthus imperialis*, which was uniaxial loaded until the first acoustic signal occurred. Initial crack propagation at the medulla-radiating layer interface is displayed in high magnification in the red box. The advanced stage of crack propagation is demonstrated on the left-hand side: the crack is deflected along the stereom sheet.

3.2. Anisotropic Ceramics

3.2.1. Bioinspired Structural Graded Ceramic

A structural graded ceramic with varying cell sizes on the μm scale is producible using the modified mold design (Figure 3B) and experimental conditions for preparing the suspension described in Section 2.2. The gradation is restricted to the lower region of the mold being directly adjacent to the bottom plate of the mold. Therefore, the distinct structural gradation is present in the first 10% of height from the ceramic. Nevertheless, the fabrication of a graded pore structure with anisotropic cell channels in the ceramic is possible in one single process step. The structural graded ceramic appearing similar to the core–shell structure of the sea urchin spine is displayed in Figure 15A. The inner center of the ceramic comprises polygonal cell channels with almost directional cell channels (parallel to the freezing direction) and is termed as 'Zone C' (Figure 15B): the pore diameters vary between 50 and 180 µm (Figure 15C). Further outwards the diameter of the polygonal cell channels becomes smaller. This area is declared as 'Zone B'. Their pore diameters are between 30 and 50 µm whereby larger pores are visible at few other areas (Figure 15A,C). The cell channels are oriented almost parallel to the freezing direction and tilt increasingly sidewards towards the outside. Very large channel sizes are present in the outermost area of the ceramic, namely in 'Zone A' (Figure 15A). Due to the great variation of the cell sizes in 'Zone A', it is not feasible to determine an approximate value for the cell sizes in this zone.

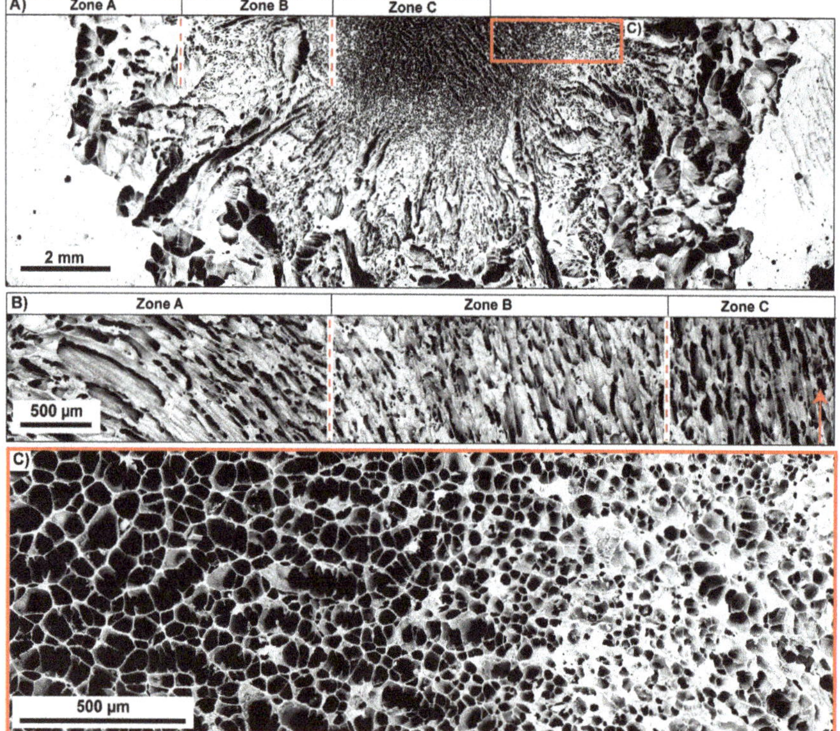

Figure 15. Structural graded ceramic inspired by the microstructure of *Phyllacanthus imperialis*. The microstructure of the ceramic and its division in different zones is displayed in (**A**) that illustrates the cross-sectional area perpendicular to the freezing direction. The orientation of the cell channels is given in (**B**). The red arrow demonstrates the freezing direction. A higher magnification of the graded structure is displayed in (**C**). The area is marked as red rectangle in (**A**).

3.2.2. Ceramics with No Gradation

The layered strut arrangement of the radiating layer and the polygonal cell arrangement of the medulla can be transferred in individual ceramics during freeze-casting using conventional cylindrical molds (Figure 3A). The pore morphology is influenced by the composition of the suspension. Both ceramic types (referred here as polygonal and oblate cells, Table 1) have the same solid loading of 14.4 vol.% in common and are, thus, characterized by an average porosity of approximately 79.5%. Different pore morphologies resembling to the spine's radiating layer and medulla can be manufactured using gelatin as additive in the water-based alumina suspension. Using a gelatin concentration of

3.5 vol.% leads to cellular oblate pore structures (perpendicular to the freezing direction; Figure 16A). The oblate cells are largely aligned in one direction. At some places, comparably smaller cells interrupt the sequence of large oblate cells, but under maintenance of the equal cell orientation in principle. Ceramic bridges are within the large oblate cells. The high anisotropic character of the cell channels can be seen in Figure 16B. Polygonal cellular cell shapes can be manufactured with a gelatin concentration of 6.8 vol.% (Figure 16C). It appears that the directionality of the cell channels becomes vaguer and obtains more a foam-like character (Figure 16D). As can be clearly seen that the configuration of the pore system, i.e., the degree of the interconnectivity of the cell walls is dependent on the gelatin concentration in the suspension.

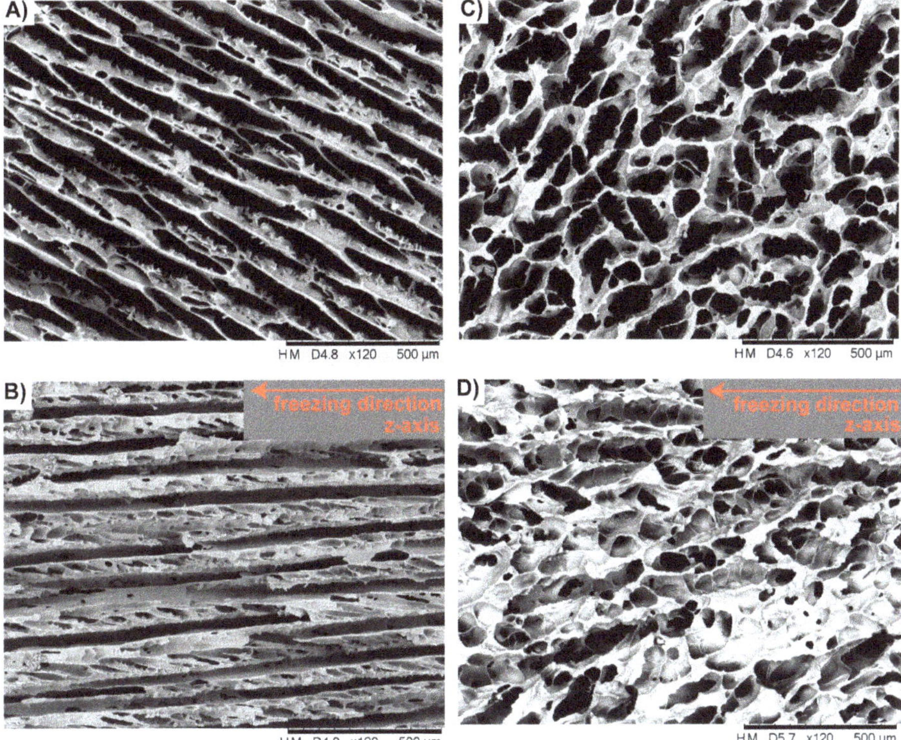

Figure 16. Freeze-casted cellular ceramics inspired by the strut arrangement of the spine's radiating layer and medulla. Both ceramic types have the solid loading of 14.4 vol.% in common and differentiate in terms of the used gelatin concentration in the suspension. A gelatin concentration of 3.5 vol.% has been used to manufacture the cellular oblate cell structure, (**A**), being highly anisotropic, (**B**), in their pore shape. Polygonal cellular cell shapes, (**C**), can be manufactured with a gelatin concentration of 6.8 vol.% showing rather a foam-like character, (**D**).

The results of the maximum compressive strength and Young's modulus of the freeze-casted alumina ceramics are summarized in Figure 17. Both ceramics are characterized by the same porosity range, but differentiate in their average values for the strength and stiffness (Figure 17A,B). The ceramics comprising polygonal cellular cell channels are characterized by an average compressive strength and Young's modulus of 1700 MPa and 37 GPa, respectively. Much lower strength and stiffness display the ceramics being characterized by the large cellular oblate cells. This ceramic type has an average value for the maximum compressive strength of 900 MPa. The average Young's modulus is 25 GPa.

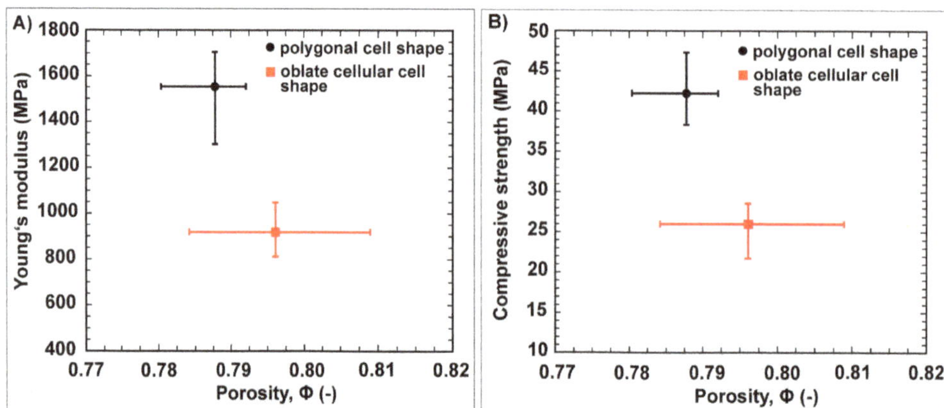

Figure 17. The plots in (**A,B**) show the average Young's modulus and maximum compressive strength of the freeze-casted ceramics as a function of the porosity, Φ.

The stress–strain curves and corresponding microphotographs of the fracture behavior of the freeze-casted ceramics are demonstrated in Figure 18. The ceramic characterized by the polygonal cellular cells compensate the stress by crumbling after the linear elastic increase (Figure 18a,b; white letters). Therefore, the structure flakes off into smaller parts starting from the upper region of the ceramic (Figure 18c,d; white letters). These two fracture mechanisms are capable to avoid the decrease of stress in large scales. By the mode of crumbling and flaking the ceramic is rather capable to keep the acting stress on a constant level. Compared to this, the ceramic with its cellular oblate anisotropic cells is characterized by a steeper gradual decrease of the compensating stress (Figure 18, red curve)). After reaching the maximum compressive strength, vertical cracks propagate through the structure intensifying with progressing compression (Figure 18a–c; red letters). This multiple propagation of vertical cracks results in a division into several lath-like segments tilting increasingly sidewards (Figure 18d,e; red letters).

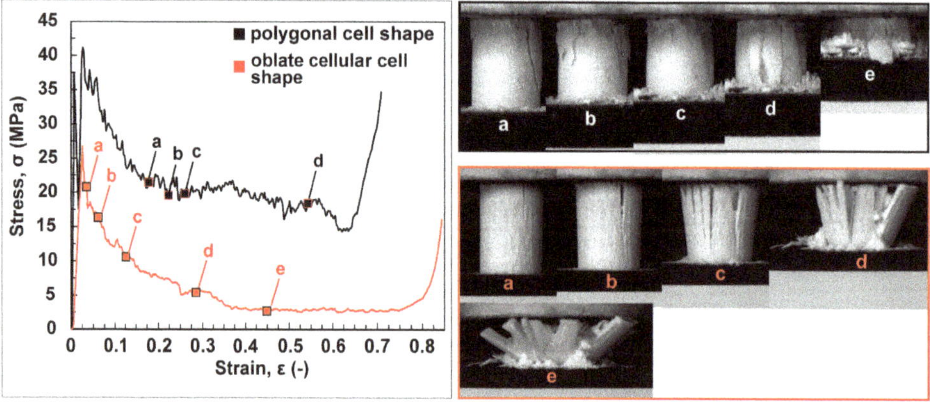

Figure 18. The stress (σ)-strain (ε) curves and the corresponding microphotographs of the fracture behavior of the freeze-casted ceramics. The ceramics including the polygonal cells compensate stress by a progressive interplay of crumbling and flaking (a–e, white letters). An intensive lath-like segmentation (a–e, red letters) can be observed for the ceramics being characterized by cellular oblate anisotropic cells.

4. Discussion

4.1. Freeze-Casted Ceramic with Gradation Features Inspired by Natural Core–Shell Systems

A modified mold design in the freeze-casting procedure allows to fabricate a structural graded ceramic being similar to concentric core–shell systems found in nature (e.g., plant stems and spine of *Phyllacanthus imperialis*). Only one single processing step is required to achieve different concentric cellular structures distinguishing according to their pore channel sizes and orientations. The differentiation in cellular substructures, anisotropy and pore channel size of the freeze-casted ceramic is similar to the lightweight construction of the spine. A comparable visual display of the spine microstructure (in particular of the interface between the medulla and radiating layer) and structural graded cellular ceramic (Zone B and C) demonstrate that the pore channel sizes are not identical (Figure 19A,B), but are very close together and have, thus, the same order of magnitude. The distribution of the porosity of the structurally graded ceramic corresponds to the cellular stereom structures of the radiating layer and medulla: both are characterized by an ascending porosity from the outside to inside.

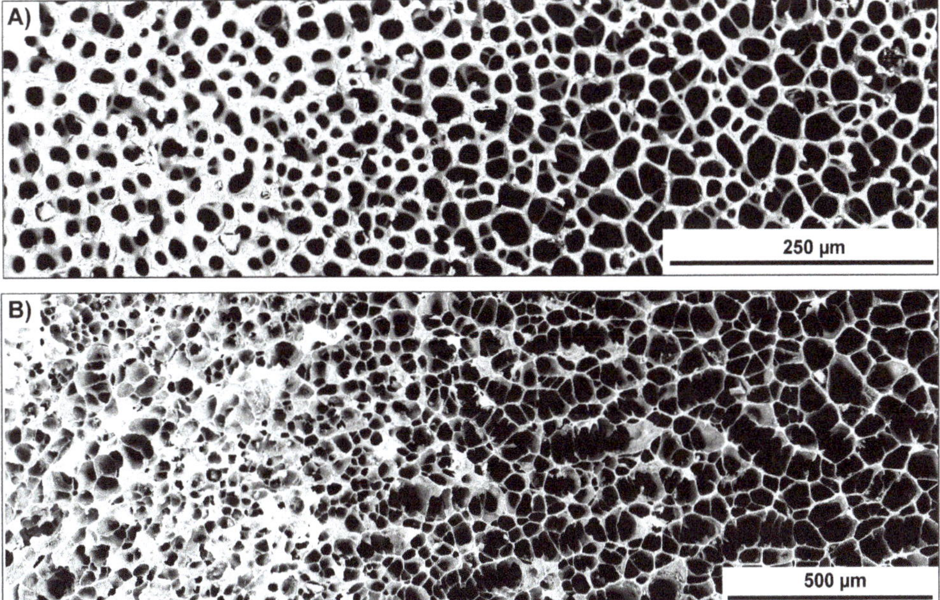

Figure 19. Comparable display of BSE images of the spine microstructure of *Phyllacanthus imperialis* and of the structural graded ceramic manufactured via freeze-casting. The cross section (perpendicular to the z-axis) of the spine microstructure is displayed in (**A**) demonstrating the radiating layer and medulla. Part of the structural graded freeze-cast ceramic with its cross section is presented in (**B**).

The polygonal cellular shaping and their directionality of the pores from the ceramic-based manufacturing concurs with the shaping and anisotropy of the pores of the sea urchin spine. The structural graded ceramic and the biomimetic concept generator share a lot of common features like the differentiation in several cellular, anisotropic structures and the polygonal cell shape. These characteristics are also present in cellular, organic core–shell systems of plant stems [7,8,47].

An implementation of a protective, but permeable external cover such as the lance sea urchin spine's cortex was only possible on a limited basis. The orientation of the pore channels agrees with the spine's cortex, but differentiate in terms of the wide distribution of the pore channel sizes. The cortex is characterized by a highly branched pore channel system with an average constant pore channel size of 17 µm. In contrast, the pore channels in the cortex-alike structure in the ceramic can obtain dimensions of >200 µm in diameter appearing as dense pore channel bundle in the microstructure. As a consequence, the quantity of solidified material between the cell channels is relatively small compared to the spine's cortex. Microstructural observations (see Figure 15, Zone A) have demonstrated that the almost dense cortex structures of the spine cannot be translated properly into a ceramic structure. The detailed wedge-like structures of the spine (i.e., the superstructure) cannot be transferred either. The fact that the cortex structure and superstructures of spine could not be transferred into a ceramic structure in their fully extent is due to the manufacturing technique itself. The subtleties of the stereom layers as well as their wedge-like enclosure of the lateral branches of the medulla cannot be depicted in detail with freeze-casting. Therefore, the superstructure has to be simplified within the framework of the possibilities of the freeze-casting method: the radial character of the pores and accompanied cell walls can be improved by lowering significantly the freezing temperatures using the mold design in Figure 3B. Significantly lower freezing temperatures enhance the ice crystal growth and its radial propagation through the suspension. The increased propagation velocity of the ice crystals allows to disrupt the discontinuous gelatin–water network more easily. Radially ordered cellular structures could be obtained in this way resembling the radially arranged stereom sheets in the sea urchin spine's radiating layer and the intermediate layers of plant stems in a simplified manner. With this processing procedure, differentiated structures similar to the cellular microstructure of the sea urchin spine's core (medulla and radiating layer) and of plant stems can be fairly transferred into a structurally graded ceramic. This one-step procedure, however, excludes the manufacturing of a similar ceramic-based structure to the spine's cortex. Since a suspension with a particular solid content is frozen, substructures can be fabricated in the ceramic that differ only minimally in porosity. The radiating layer and the medulla are characterized by a porosity of 76.6 ± 3.0 and 85.7 ± 2.8%, respectively. The difference in the porosities between them is comparatively small and is realizable in one single ceramic via freeze-casting. In contrast, the difference in the porosities between the elements of the spine's core and the cortex is large. In order to realize an external cover such as the spine's cortex in the ceramic, a second radial freezing step is required involving a suspension with a higher solid content.

4.2. Freeze-Casted Ceramics with Anisotropic Cellular Structures

The manufacturing of a porous ceramic with various gradation features is essentially dependent on the mold design in the freeze-casting process. Fabrication of individual anisotropic cellular structures in the ceramic involving no gradation features at all is rather a function of the composition of the suspension and freezing condition. By the use of gelatin as additive in the water-based alumina suspension, cellular pores systems on the µm scale can be manufactured resembling structures found in nature. Depending on the gelatin concentration in the suspension anisotropic polygonal channels (see Table 1) and cellular oblate pore systems (see Table 1) are producible that appear very similar to the spine's interior, i.e., the medulla and radiating layer, respectively. These ceramic structures, in particular the anisotropic polygonal pore systems look also similar to other cellular materials in plants such as the cross section of a milkweed stem [48] and a porcupine quill [7]. Freeze-casting is, therefore, a practicable manufacturing technique to create a multitude of anisotropic plant-alike cell structures in ceramics.

The microcellular ceramics shown here differ in terms of their strength and stiffness, even though they have similar porosities. Therefore, the strength and stiffness of the ceramics strongly relate to the shaping of the pore system. The cross sections in Figure 20 (2D view, perpendicular to the freezing direction: complete filling of the cavities with epoxy resin) reveal that the cell walls are comparatively stronger interconnected due to the polygonal pore shape in the ceramic. Larger stiffness and strength are a result of the high interconnectivity of the cell walls (Figure 20A). In contrast, several cell walls are isolated in space and, thus, unconnected in the cellular oblate cell system

(Figure 20B). Hence, comparatively lower values in the strength and stiffness are the result of the reduced interconnectivity in the oblate cell system. The fracture behavior of the ceramics under uniaxial compression is likewise strongly related to the shaping of the cell system: a progressive interplay of crumbling and flaking of small lateral pieces was observed for the ceramics with the polygonal anisotropic cell channels; the segmental division into several lath-like segments caused by a multitude of vertical cracks was a characteristic for the ceramics comprising the oblate cellular cell system. Since several cell walls are isolated in space in this latter ceramic type, the cell walls are subjected to high shear stresses and bending moments and, therefore, susceptible to high deflection during uniaxial compression. Consequently, large segments can spall off from the structure. The structural integrity is no longer ensured and the capability to absorb stress is, thus, reduced greatly. The greatly linked anisotropic polygonal cell architecture prevents shearing and bending moments of the cell walls. The stiffing, therefore, increases the mechanical stability and keeps the material loss comparatively small. It implies that the major quantity of the ceramic tends to retain its structural integrity. Consequently, a high quantity of stress is, therefore, required to break the strong cohesive structure. Both main fracture modes in the ceramics have the structural segmentation in common. They differentiate in terms of the intensity of the structural segmentation or destruction being comparable to the mode of 'quasi brittle' fracturing of the sea urchin spine.

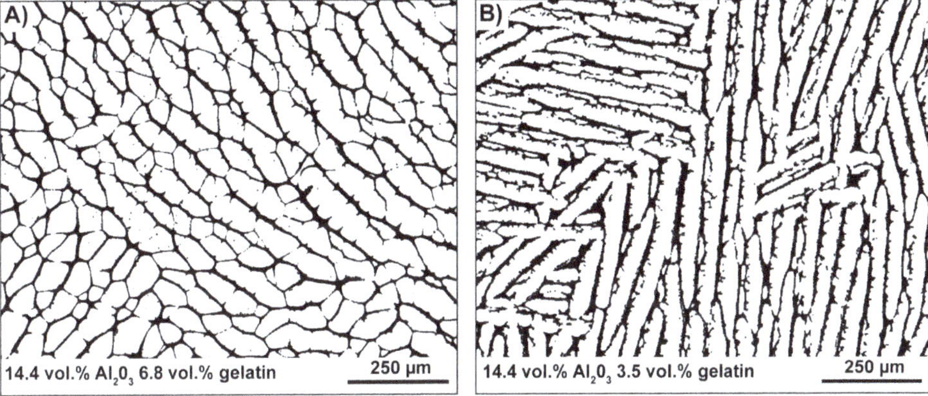

Figure 20. BSE images of the cross sections of the freeze-casted ceramics (perpendicular to the freezing-direction). The cavities of the ceramics were filled with epoxy resin (=white areas). Therefore, the varying degrees of interconnectivity of the struts in the ceramics is demonstrated in (**A**,**B**) depending on the gelatin concentration.

There is a structural difference between microstructure of the sea urchin spine and the cellular ceramics presented here. The key element of the structural segmentation for the freeze-casted ceramics is dependent on the degree of interconnectivity of the cell walls. In contrast, the structural segmentation or destruction of the sea urchin spine is dependent on the core's wedge-shaped interlocking of the microstructure and of the quantity of the surrounding cortex.

4.3. The Plant Stem-Alike Core–Shell Construction in the Lance Sea Urchin Spines

The core of the spine with its cellular substructures (i.e., the medulla and radiating layer) is characterized by a highly interconnected complex stereom network: two different stereom types are united in a special arranged wedge-shaped configuration according to the key–lock principle (i.e., a superstructure). The intelligent wedge-alike interlocking saves building material and deflect cracks advantageously. Uniaxial compression experiments on the core have demonstrated that the stress is concentrated at the interface between the medulla and radiating layer. Studies on natural core–shell systems of plant stems, animal quills and bird feather rachis under uniaxial compression [7,44,45] have shown that the stress is also maximized at the interface and decays radially inward. So, the first cracks appear at the interface and run along the stereom layers to the exterior of the core. Therefore, lath-like segments emerge in the spine structure. The size of the separated segments could be limited by the wedges in principle to keep the material loss as low as possible. With the minimal use of building material, the core of the spine is optimized in terms of the ability to provide stiffness, strength and an advantageous energy dissipation. In addition, the inner complex mesh network is surrounded by the permeable shell structure—the cortex.

The quantity of the cortex in the spine segment dictates in the main the dimension of the structural integrity and the resilience of the spine segment to mechanical stress. A high-volume fraction of the cortex at the spine tip (a/t ratio > 9.2), which decreases towards to the spine base (a/t ratio < 9.2), may indicate a possible adaption of the sea urchin to its natural habitat. The sea urchin wedges themselves during the day time in the reef cavity in order to protect themselves against the high hydrodynamic environment and predators [49,50]. The spine tip is directly in contact with the hard substrate due to the wedging. A high resilience at the spine tip is, therefore, required, since a high punctual load, which lasts for hours, act on the tip of the spine. The high-volume fraction of the cortex at the spine tip is, thus, useful to enhance the resilience in principle. The 'quasi bulging' of the cortex (failure mode II) might be an indication for a reduced rigidity. A solid solution hardening of the material is achieved by the magnesium incorporation in the crystal structure [51]. Investigations have shown that the average magnesium concentration in the cortex is 5.1 mole percent $MgCO_3$ [52]. This is significantly lower than the concentrations observed in the medulla (5.9 mole percent $MgCO_3$) and the radiating layer (8.1–0.5 mole percent $MgCO_3$). A decrease of the magnesium in the cortex reduces also the degree of brittleness, which allows, therefore, minimal elastic reactions. These minimal elastic reactions are expressed in pure compressive loading of the spine segments as 'quasi bulging'. A minimal degree of elasticity of the cortex at the spine tip is required, since the sea urchin has to modify the pressure for the wedging according to the prevailing hydrodynamic conditions. Changing flow velocities and the occurrence of predators imply a highly dynamic environment for the sea urchin. The pressure acting on the spine is constantly readjusted to the changes of the environment. Therefore, the punctual load at the spine tip fluctuates during the wedging. A rigid cortex would directly lead to the actual material fracture of the upper spine segment. This interpretation of the results extends the rather chemical function of the cortex given by [52]. They have argued that the low magnesium concentration in the cortex compared to the inner core and the increased density of the cortex is responsible for the large acidification resistance. Probably the large acidification resistance enabled them to survive the Permian-Triassic crisis, which was associated with severe acidification events of the ocean. Segments from the lower half of the spine are characterized by a rigid cortex–core connection. Therefore, these spine segments keep widely the structural integrity under high compressive loading and maintain the degree of segmentation as low as possible. Conversely, segments from the spine tip are strongly subjected to structural segmentation, since the cortex was spalled off right at the beginning of the experiment. Therefore, the core was not stabilized anymore by the cortex. The lower half of the spine has been optimized towards keeping further material losses as low as possible, when the upper half of the spine does not exist anymore.

The level of compressive loading acting on the sea urchin spine segment in the experimental setting (>500 N) was significantly higher than the mechanical loading conditions prevailing in the natural habitat of the sea urchin. Such compressive loading conditions in the experimental setting would probably press the spine into the test. The test would fracture, thus, catastrophically before the spine would dissipate energy. Therefore, the ability to dissipate energy advantageously is not necessarily an adaption to its mechanical environment, but rather a positive by-product of the porous, hierarchically-organized material by itself. Nevertheless, the results of the mechanical testing of the spine segment offered room to discuss the function of the structural elements of the spine in principle.

5. Conclusions

The manufacturing of a porous ceramic with several gradation features inspired by natural cylindrical core–shell systems is possible using freeze-casting as processing technique. Using a special mold design and optimized suspension for the freeze-casting technique allows to achieve a differentiation in several cellular substructures, anisotropy and pore channel sizes in the ceramic comparable to the microstructure of plant stems and the spine of *Phyllacanthus imperialis*. Biological structures such as the permeable cortex of spine and the regular arrangement of the stereom sheets (=radiating layer) were only transferred into a ceramic structure to a limited degree. Several substructures can be fabricated in one single ceramic that differ only minimally in the porosity, since a suspension with a particular solid content is used for the freezing process. Therefore, structures such as the spine's medulla and radiating layer are manufacturable in an abstracted way in ceramics using one single processing step, because the difference in the porosities between them is very small (approximately 10%). However, for realizing a cortex-alike structure in the ceramic, a second radial freezing step is required involving a suspension with a higher solid content.

Ceramics with individual cellular pore systems (i.e., no gradation features) are possible using a conventional mold design during the freeze-casting process. Ceramic structures being similar to cellular materials in plants such as milkweed and porcupine stem can be manufactured using gelatin as additive in the water-based suspension. Starting at a gelatin concentration of 6.8 vol.% with a gradual reduction of the gelatin concentration by 50% has shown that polygonal and cellular oblate pore systems can be manufactured in this way. A reduced lamellar spacing through the use of gelatin is an important structural factor that contributes to the improvement of mechanical properties. The structural modification through gelatin brings the overall structure closer to that of a honeycomb using its improved mechanical efficiency. The increased number of stiffening and strengthening bridges stabilizes the overall structure through rib stiffening and reduce failure due to shearing.

The new results focusing on the relation between the material properties and the specific microstructural configuration of the spine offer room for discussion in the light of the mechanical environment of the sea urchin *Phyllacanthus imperialis*. The wedge-shaped interlocking of two different stereom meshes according to the key–lock principle of the spine's inner microstructure is a biological optimized structure to provide stiffness, strength and a beneficial energy dissipation with the minimal use of material. The surrounding permeable shell structure, the cortex, dictates the dimension of the structural integrity and the resilience of the spine to mechanical stress. A high-volume fraction of the cortex at the spine tip enhances the mechanical stability and is required, since the sea urchins wedge themselves between the reef cavities during the day time. A high resilience at the spine tip is, therefore, required to withstand the high punctual loadings, which can last for hours. The cortex is characterized by lower magnesium concentrations compared to the inner structural elements of the spine. Low magnesium concentrations reduce the rigidity of the cortex and allow minimal elastic reaction that was experimentally visible under extreme loading conditions of the spine segment ('quasi bulging'). A minimal degree of elasticity of the cortex at the spine tip is reasonable, since the sea urchin have to modify the pressure for the wedging according to the prevailing hydrodynamic conditions. In contrast, segments from the lower half of the spine are characterized by a rigid cortex-core connection. Thus, the lower half of the spine has been optimized towards minimizing further material losses, when the upper half of the spine has been broken off.

Porous anisotropic ceramics are widely used as catalyst supports and as particulate filters for vehicular emission control. New geometries and materials are permanently developed for both mobile and stationary applications. As a result, considerable work is done studying higher cell density and thinner wall configurations. A higher cell density and thinner cell walls of the substrates increase the geometric surface area and reduce the hydrocarbon emissions during all phases of the catalytic conversion. To avoid a decreasing substrate strength due to thinner ceramic walls, the special key–lock arrangement of the wedges and the permeable shell-like construction from the lance sea urchin spine can be integrated in the porous filter system. Hereby, a mechanical strengthening of the overall structure is possible while maintaining the necessary flow properties due to the anisotropic pore morphologies.

Author Contributions: Conceptualization, K.K. and K.G.N.; methodology, K.K.; software, K.K.; validation, K.K. and K.G.N.; formal analysis, K.K.; investigation, K.K.; resources, K.G.N.; data curation, K.K.; writing—original draft preparation, K.K.; writing—review and editing, K.K. and K.G.N.; visualization, K.K.; supervision, K.G.N.; project administration, K.G.N.; funding acquisition, K.G.N. All authors have read and agreed to the published version of the manuscript.

Funding: This work was funded by the German Research Foundation (DFG) as part of the Transregional Collaborative Research Centre (SFB-TRR) 141 'Biological design and integrative structures' (project A02 'Plants and animals as source of inspiration for energy dissipation in load bearing systems and facades').

Institutional Review Board Statement: Not applicable.

Informed Consent Statement: Not applicable.

Data Availability Statement: Data is contained within the article.

Acknowledgments: The authors gratefully acknowledge Hermann Finckh for performing the μCT measurements at the Institute of Textile Technology and Process Engineering (ITV) Denkendorf. The authors would like to thank Simone Schafflick and Barbara Maier for the preparational help. We thank the editors, reviewers and Michael Bergler for their constructive comments which improved the manuscript.

Conflicts of Interest: The authors declare no conflict of interest.

References

1. Schmier, S.; Hosoda, N.; Speck, T. Hierarchical structure of the Cocos nucifera (coconut) endocarp: Functional morphology and its influence on fracture toughness. *Molecules* **2020**, *25*, 223. [CrossRef] [PubMed]
2. Schmier, S.; Jentzsch, M.; Speck, T.; Thielen, M. Fracture mechanics of the endocarp of Cocos nucifera. *Mater. Des.* **2020**, *195*, 108944. [CrossRef]
3. Bauer, G.; Speck, T.; Blömer, J.; Bertling, J.; Speck, O. Insulation capability of the bark of trees with different fire adaptation. *J Mater. Sci.* **2010**, *45*, 5950–5959. [CrossRef]
4. Born, L.; Körner, A.; Schieber, G.; Westermeier, A.S.; Poppinga, S.; Sachse, R.; Bergmann, P.; Betz, O.; Bischoff, M.; Speck, T.; et al. *Fiber-Reinforced Plastics with Locally Adapted Stiffness for Bio-Inspired Hingeless, Deployable Architectural Systems*, 1st ed.; Trans Tech Publications Ltd.: Zurich, Switzerland, 2017; pp. 689–696. [CrossRef]
5. Grun, T.B.; von Scheven, M.; Bischoff, M.; Nebelsick, J.H. Structural stress response of segmented natural shells: A numerical case study on the clypeasteroid echinoid Echinocyamus pusillus. *J. R. Soc. Interface* **2018**, *15*, 20180164. [CrossRef] [PubMed]
6. Kovaleva, D.; Gericke, O.; Kappes, J.; Haase, W. Rosenstein-Pavilion: Towards resource efficiency by design. *Beton Stahlbetonbau* **2018**, *113*, 433–442. [CrossRef]
7. Karam, G.N.; Gibson, L.J. Biomimicking of animal quills and plant stems: Natural cylindrical shells with foam cores. *Mater. Sci. Eng. C* **1994**, *2*, 113–132. [CrossRef]
8. Schott, R.T.; Roth-Nebelsick, A. Ice nucleation in stems of trees and shrubs with different frost resistance. *IAWA J.* **2018**, *39*, 177–190. [CrossRef]
9. Smith, A.B. Stereom Microstructure of the Echinoid Test. *Spec. Pap. Palaeontol.* **1980**, *25*, 1–324.
10. Smith, A.B. Biomineralization in Echinoderms. In *Skeletal Biomineralization: Patterns, Processes and Evolutionary Trends*, 1st ed.; Carter, J.G., Ed.; VanNostrand Reinhold: New York, NY, USA, 1990; pp. 413–443.
11. Märkel, K.; Kubanek, F.; Willgallis, A. Polycristalline calcite in sea urchins. *Z. Zellforsch* **1971**, *119*, 355–377. [CrossRef]
12. Cölfen, H.; Antonietti, M. *Mesocrystals and Nonclassical Crystallization*; John Wiley and Sons: Hoboken, NJ, USA, 2008. [CrossRef]
13. Seto, J.; Zhang, Y.; Hamilton, P.; Wilt, F. The localization of occluded matrix proteins in calcareous spicules of sea urchin larvae. *J. Struct. Biol.* **2004**, *148*, 123–130. [CrossRef]
14. Presser, V.; Schultheiß, S.; Berthold, C.; Nickel, K.G. Sea Urchin Spines as a Model-System for Permeable, Light-Weight Ceramics with Graceful Failure Behavior. Part, I. Mechanical Behavior of Sea Urchin Spines under Compression. *J. Bionic Eng.* **2009**, *6*, 203–213. [CrossRef]
15. Presser, V.; Schultheiß, S.; Kohler, C.; Berthold, C.; Nickel, K.G.; Vohrer, A.; Finckh, H.; Stegmaier, T. Lessons from nature for the construction of ceramic cellular materials for superior energy absorption. *Adv. Eng. Mater.* **2011**, *13*, 1042–1049. [CrossRef]

16. Klang, K.; Bauer, G.; Toader, N.; Lauer, C.; Termin, K.; Schmier, S.; Kovaleva, D.; Haase, W.; Berthold, C.; Nickel, K.G.; et al. Plants and Animals as Source of Inspiration for Energy Dissipation in Load Bearing Systems and Facades. In *Biomimetic Research for Architecture and Building Construction: Biological Design and Integrative Structures*, 1st ed.; Knippers, J., Nickel, K.G., Speck, T., Eds.; Springer International Publishing: Cham, Switzerland, 2016; pp. 109–133.

17. Presser, V.; Kohler, C.; Živcová, Z.; Berthold, C.; Nickel, K.G.; Schultheiß, S.; Gregorová, E.; Pabst, W. Sea Urchin Spines as a Model-System for Permeable, Light-Weight Ceramics with Graceful Failure Behavior. Part II. Mechanical Behavior of Sea Urchin Spine Inspired Porous Aluminum Oxide Ceramics under Compression. *J. Bionic Eng.* **2009**, *6*, 357–364. [CrossRef]

18. Tang, Z.; Kotov, N.A.; Magonov, S.; Ozturk, B. Nanostructured artificial nacre. *Nat. Mater.* **2003**, *2*, 413–418. [CrossRef]

19. Podsiadlo, P.; Kaushik, A.K.; Arruda, E.M.; Waas, A.M.; Shim, B.S.; Xu, J.; Nandivada, H.; Pumplin, B.G.; Lahann, J.; Ramamoorthy, A.; et al. Ultrastrong and stiff layered polymer nanocomposites. *Science* **2007**, *318*, 80–83. [CrossRef]

20. Yao, H.-B.; Fang, H.-Y.; Tan, Z.-H.; Wu, L.-H.; Yu, S.-H. Biologically inspired, strong, transparent, and functional layered organic-inorganic hybrid films. *Angew. Chem. Int. Ed.* **2010**, *49*, 2140–2145. [CrossRef]

21. Cheng, Q.; Wu, M.; Li, M.; Jiang, L.; Tang, Z. Ultratough artificial nacre based on conjugated cross-linked graphene oxide. *Angew. Chem. Int. Ed.* **2013**, *52*, 3750–3755. [CrossRef]

22. Zhang, M.; Wang, Y.; Huang, L.; Xu, Z.; Li, C.; Shi, G. Multifunctional Pristine Chemically Modified Graphene Films as Strong as Stainless Steel. *Adv. Mater.* **2015**, *27*, 6708–6713. [CrossRef]

23. Chavez-Valdez, A.; Shaffer, M.; Boccaccini, A.R. Applications of graphene electrophoretic deposition. A review. *J. Phys. Chem. B* **2013**, *117*, 1502–1515. [CrossRef] [PubMed]

24. Fukasawa, T.; Ando, M.; Ohji, T.; Kanzaki, S. Synthesis of Porous Ceramics with Complex Pore Structure by Freeze-Dry Processing. *J. Am. Ceram. Soc.* **2001**, *84*, 230–232. [CrossRef]

25. Deville, S. Freeze-casting of porous ceramics: A review of current achievements and issues. *Adv. Eng. Mater.* **2008**, *10*, 155–169. [CrossRef]

26. Deville, S. Freeze-casting of porous biomaterials: Structure, properties and opportunities. *Materials* **2010**, *3*, 1913–1927. [CrossRef]

27. Peppin, S.; Elliott, J.; Worster, M.G. Solidification of colloidal suspensions. *J. Fluid Mech.* **2006**, *554*, 147–166. [CrossRef]

28. Peppin, S.; Worster, M.G.; Wettlaufer, J.S. Morphological instability in freezing colloidal suspensions. *Proc. R. Soc. A Math. Phys. Eng. Sci.* **2007**, *463*, 723–733. [CrossRef]

29. Peppin, S.; Wettlaufer, J.S.; Worster, M.G. Experimental verification of morphological instability in freezing aqueous colloidal suspensions. *Phys. Rev. Lett.* **2008**, *100*, 238301. [CrossRef]

30. Hunger, P.M.; Donius, A.E.; Wegst, U. Structure-property-processing correlations in freeze-cast composite scaffolds. *Acta Biomater.* **2013**, *9*, 6338–6348. [CrossRef] [PubMed]

31. Lichtner, A.; Roussel, D.; Jauffrès, D.; Martin, C.L.; Bordia, R.K. Effect of Macropore Anisotropy on the Mechanical Response of Hierarchically Porous Ceramics. *J. Am. Ceram. Soc.* **2016**, *99*, 979–987. [CrossRef]

32. Liu, X.; Rahaman, M.N.; Fu, Q.; Tomsia, A.P. Porous and strong bioactive glass (13-93) scaffolds prepared by unidirectional freezing of camphene-based suspensions. *Acta Biomater.* **2012**, *8*, 415–423. [CrossRef]

33. Munch, E.; Saiz, E.; Tomsia, A.P.; Deville, S. Architectural control of freeze-cast ceramics through additives and templating. *J. Am. Ceram. Soc.* **2009**, *92*, 1534–1539. [CrossRef]

34. Naviroj, M.; Voorhees, P.W.; Faber, K.T. Suspension- and solution-based freeze casting for porous ceramics. *J. Mater. Res.* **2017**, *32*, 3372–3382. [CrossRef]

35. Fukushima, M.; Yoshizawa, Y.-I.; Ohji, T. Macroporous ceramics by gelation-freezing route using gelatin. *Adv. Eng. Mater.* **2014**, *16*, 607–620. [CrossRef]

36. Speck, O.; Speck, D.; Horn, R.; Gantner, J.; Sedlbauer, K.P. Biomimetic bio-inspired biomorph sustainable? An attempt to classify and clarify biology-derived technical developments. *Bioinspir. Biomimetics* **2017**, *12*, 11004. [CrossRef] [PubMed]

37. de Lamarck, J.B.M. *Histoire Naturelle des Animaux sans Vertébres, Tome Deuxiéme*; Bailliére Libraire: Paris, France, 1836; p. 568.

38. DeFoe, O.K.; Compton, A.H. The density of rock salt and calcite. *Phys. Rev.* **1925**, *25*, 618–620. [CrossRef]

39. Sun, Z.; Li, B.; Hu, P.; Ding, F.; Yuan, F. Alumina ceramics with uniform grains prepared from Al_2O_3 nanospheres. *J. Alloy. Compd.* **2016**, *688*, 933–938. [CrossRef]

40. Hautcoeur, D.; Gonon, M.; Baudin, C.; Lardot, V.; Leriche, A.; Cambier, F. Alumina Porous Ceramics Obtained by Freeze Casting: Structure and Mechanical Behaviour under Compression. *Ceramics* **2018**, *1*, 8. [CrossRef]

41. Ridler, T.W.; Calvard, S. Picture thresholding using an iterative selectction methode. *IEEE Trans. Syst. Man. Cybern.* **1978**, *8*, 630–632. [CrossRef]

42. Doube, M.; Klosowski, M.M.; Arganda-Carreras, I.; CordeliÃ¨res, F.P.; Dougherty, R.P.; Jackson, J.S.; Schmid, B.; Hutchinson, J.R.; Shefelbine, S.J. BoneJ: Free and extensible bone image analysis in ImageJ. *Bone* **2010**, *47*, 1076–1079. [CrossRef]

43. Dawson, M.A.; Gibson, L.J. Optimization of cylindrical shells with compliant cores. *Int. J. Solids Struct.* **2007**, *44*, 1145–1160. [CrossRef]

44. Karam, G.N.; Gibson, L.J. Elastic buckling of cylindrical shells with elastic cores-I. Analysis. *Int. J. Solids Struct.* **1995**, *32*, 1259–1283. [CrossRef]

45. Karam, G.N.; Gibson, L.J. Elastic buckling of cylindrical shells with elastic cores-II. Experiments. *Int. J. Solids Struct.* **1995**, *32*, 1285–1306. [CrossRef]

46. Yu, S.; Liu, J.; Wei, M.; Luo, Y.; Zhu, X.; Liu, Y. Compressive property and energy absorption characteristic of open-cell ZA22 foams. *Mater. Des.* **2009**, *30*, 87–90. [CrossRef]
47. Speck, T.; Burgert, I. Plant Stems: Functional Design and Mechanics. *Annu. Rev. Mater. Res.* **2011**, *41*, 169–193. [CrossRef]
48. Gibson, L.J.; Ashby, M.F. *Cellular Solids: Structure and Properties*, 2nd ed.; Cambridge University Press: Cambridge, UK, 2014. [CrossRef]
49. Nebelsick, J.H. Echinoid distribution by fragment identification in the northern Bay of Safaga, Red Sea, Egypt. *Palaios* **1992**, *7*, 316–328. [CrossRef]
50. Nebelsick, J.H. Biodiversity of shallow-water red sea echinoids: Implications for the fossil record. *J. Mar. Biol. Assoc. UK* **1996**, *76*, 185–194. [CrossRef]
51. Magdans, U.; Gies, H. Single crystal structure analysis of sea urchin spine calcites: Systematic investigations of the Ca/Mg distribution as a function habitat of the sea urchin and the sample location in the spine. *Eur. J. Mineral.* **2004**, *16*, 261–268. [CrossRef]
52. Dery, A.; Guibourt, V.; Catarino, A.I.; Compère, P.; Dubois, P. Properties, morphogenesis, and effect of acidification on spines of the cidaroid sea urchin Phyllacanthus imperialis. *Invertebr. Biol.* **2014**, *133*, 188–199. [CrossRef]

 *biomimetics*

Article

3D Reticulated Actuator Inspired by Plant Up-Righting Movement Through a Cortical Fiber Network

Tom Masselter [1,2,*], **Olga Speck** [1,3] and **Thomas Speck** [1,2,3]

1   Plant Biomechanics Group @ Botanic Garden Freiburg, University of Freiburg, Schänzlestraße 1,
    D-79104 Freiburg, Germany; olga.speck@biologie.uni-freiburg.de (O.S.);
    thomas.speck@biologie.uni-freiburg.de (T.S.)
2   Freiburg Materials Research Center (FMF), University of Freiburg, Stefan-Meier-Straße 21,
    D-79104 Freiburg, Germany
3   Cluster of Excellence livMatS @ FIT—Freiburg Center for Interactive Materials and Bioinspired Technologies,
    University of Freiburg, Georges-Köhler-Allee 105, D-79110 Freiburg, Germany
*   Correspondence: tom.masselter@biologie.uni-freiburg.de

Abstract: Since most plant movements take place through an interplay of elastic deformation and strengthening tissues, they are thus ideal concept generators for biomimetic hingeless actuators. In the framework of a biomimetic biology push process, we present the transfer of the functional movement principles of hollow tubular geometries that are surrounded by a net-like structure. Our plant models are the recent genera *Ochroma* (balsa) and *Carica* (papaya) as well as the fossil seed fern *Lyginopteris oldhamia*, which hold a net of macroscopic fiber structures enveloping the whole trunk. Asymmetries in these fiber nets, which are specifically caused by asymmetric growth of the secondary wood, enable the up-righting of inclined *Ochroma* and *Carica* stems. In a tubular net-like structure, the fiber angles play a crucial role in stress–strain relationships. When braided tubes are subjected to internal pressure, they become shorter and thicker if the fiber angle is greater than 54.7°. However, if the fiber angle is less than 54.7°, they become longer and thinner. In this article, we use straightforward functional demonstrators to show how insights into functional principles from living nature can be transferred into plant-inspired actuators with linear or asymmetric deformation.

Keywords: actuator; bark; biomimetics; curving demonstrators

Citation: Masselter, T.; Speck, O.; Speck, T. 3D Reticulated Actuator Inspired by Plant Up-Righting Movement Through a Cortical Fiber Network. *Biomimetics* **2021**, *6*, 33. https://doi.org/10.3390/ biomimetics6020033

Academic Editor: Matt McHenry

Received: 26 April 2021
Accepted: 21 May 2021
Published: 27 May 2021

Publisher's Note: MDPI stays neutral with regard to jurisdictional claims in published maps and institutional affiliations.

1. Introduction

Most plants lack motility and are thus bound to their location. Nevertheless, many plants are capable of complex movements, such as the snapping action in both the carnivorous Venus flytrap and the water trap plant [1,2]. However, we often do not notice plant movements, either because the movements are too fast for the human eye, such as the trapping motion of the aquatic carnivorous bladderworts *Utricularia* spp. [3,4], or because plant motion takes place over a long period of time, such as the curving of branches caused by the formation of reaction wood in trees [5,6]. During biological evolution, plants have developed a variety of movement types and actuation mechanisms that enable either a rapid response to a stimulus or a long-term adaptation to changing environmental conditions and that provide great selective advantages for the survival of the respective plant species. Since most plant movements, in contrast to those of animals, take place without muscles and conventional (localized) hinges but through an interplay of elastic deformation and strengthening tissues, they are thus ideal concept generators for biomimetic hingeless actuators [7].

1.1. Movements of Plants

In multicellular plants, the duration of a movement is a function of the smallest macroscopic dimension of the moving part. Based on poroelastic time (=fastest tissue swelling) and inertial time (=fastest tissue elastic motion), the driving forces behind plant movements can be classified into hydraulic processes and mechanical instabilities. Hydraulically driven movements include irreversible growth processes and movement caused by reversible swelling or shrinking attributable to osmotic or humidity changes. Slow hydraulic movements can be speeded up by mechanical instabilities such as the release of stored elastic energy and rapid geometric changes [8,9].

Whereas some of the movements that occur with a change in humidity or turgor are dependent on the shape of the moving organ (cf. [2,9,10]), others are closely coupled to cylindrical geometries surrounded by a net-like structure [11,12]. An example of the latter is the structural setup enabling the movement of the feet of sea urchins [13,14].

1.2. Plant-Inspired Actuators

The transfer of principles of plant movements to biomimetic technical structures is challenging, as relatively few technical components are currently available that allow the implementation of hingeless elastic opening and closure and other functional aspects typical of plants. Nevertheless, biomimetic solutions to such technical problems show great potential and should lead to jump innovations. This also holds true for bio-inspired actuators that can be used in a wide range of technological applications including elastic systems in architecture, such as those achieved for the plant-inspired façade shading systems Flectofin and Flectofold (cf. [7,15]). In a biology push process (=biomimetic bottom-up approach), the Universities of Freiburg and Stuttgart have cooperated in the development of a hygroscopic demonstrator as a proof of concept for mimicking the opening and closure of pine cones [10,16]. Plant-inspired motion ideas have also been integrated in a hygroscopically actuated exhibit (Hygroscope, Centre Pompidou) and the Hygroskin pavilion at the FRAC center in Orléans [17,18].

Similar challenges have been faced within a biomimetic technology pull process (=biomimetic top-down approach) by [19] during the development of a plant-inspired actuator without articulated hinges on an architectural scale. The authors used the opening and closing movement of grass leaves as a model for the development of a biomimetic cellular actuator. Leaf halves of *Sesleria nitida* show pronounced kinetic amplification actuated by turgor changes of fan-shaped groups of bulliform cells that lie in the upper epidermis to the right and left of the leaf midrib. Sclerenchyma strands, together with vascular bundles, form strengthening spacers between the upper and lower epidermis and an elastic hinge-like structure at the midrib. The leaf halves fold shut when the turgor pressure in the bulliform cells is low and unfold again when the bulliform cells become fully turgescent. Based on this functional principle, Mader et al. [19] developed a pneumatic cellular actuator consisting of a row of single cells with compliant hinges positioned on a plate. When the pneumatic pressure applied to each individual cell is increased, the cell becomes wider along its upper side, and the entire structure bends. A first prototype was successfully integrated into the façade shading system Flectofold [20], in which the bending of the midrib controls the hoisting of its wings.

1.3. Conventional Technical Solutions

The movements of cross-linked net-like tubular tissues that are a distinct characteristic in many plants and animals (as described above) are in current technical use, e.g., in the so-called fluidic muscle that is produced by Festo AG & Co. KG (Figure 1) [21] and in which increasing air pressure leads to a shortening of the structure. By shortening these actuators, loads can be moved with a smooth motion pattern. The explanation of this principle of action has led to a popular science publication in which the mode of operation is illustrated with a simple experiment using balloons, commercially available vegetable packaging nets, and cable ties [22]. Nevertheless, we must not assume that the fluidic muscle is a biomimetic product. It represents an analogy of form and function in nature and technology, i.e., no transfer of knowledge occurred from a natural model to the technology involved in the development of the fluidic muscle. However, the combination of fluidic muscles enables complex actuation patterns such as bending and curving and is thus able to attain high degrees of freedom, as exhibited by the trunk of an elephant [23,24].

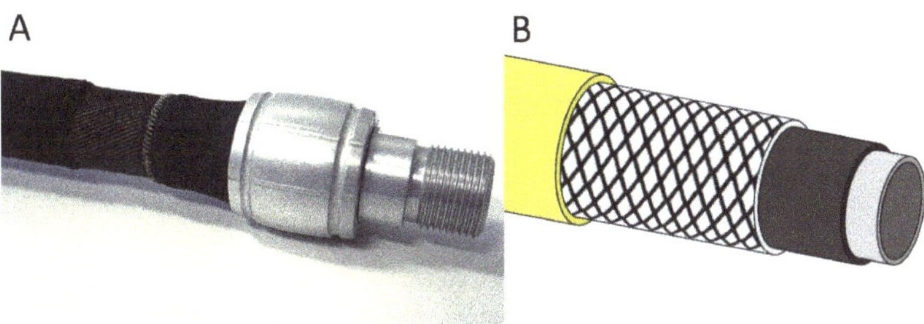

Figure 1. (**A**) The fluidic muscle (pneumatic muscle), which consists of a hollow elastomer cylinder with embedded aramid fibers, represents a tensile actuator. To make the fiber network visible, the outer coating was partially removed (© Festo AG & Co. KG, Esslingen, Germany). (**B**) Schematic drawing of a pressure-resistant hose with reinforcing fibers having an angle between the longitudinal axis and fibers of approximately 55°.

1.4. Aim of the Study

In this study, we aim to describe the development of a plant-inspired actuator that can be used to erect or bend rod-shaped objects. We describe the idea flow from the initial biological insights to various straightforward physical demonstrator models by using the biology push process (=biomimetic bottom-up approach) [25,26]. We present the six consecutive steps of the biomimetic development approach (Figure 2): (i) the biological question regarding net-like structures in plants, (ii) the selection and investigation of suitable models, (iii) the identification of the functional principle underlying the plant movement, (iv) the abstraction and translation into common language for natural scientists and engineers, (v) feasibility studies, and (vi) the outlook for possible applications in technology, education, and training.

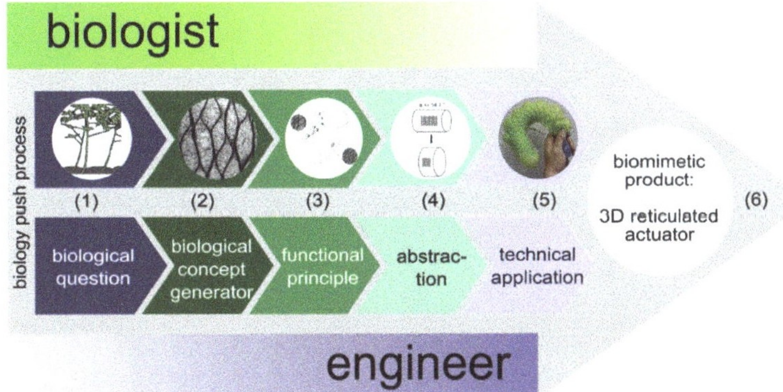

Figure 2. The biology push process (biomimetic bottom-up approach) of the curved 3D reticulated actuator. (**1**) What is the basis of the up-righting movement of inclined plant axes? (**2**) Morphological–anatomical investigations of *Lyginopteris oldhamia* revealed net-like structures in the cortex; (**3**) functional analysis of the actuation principle in stem up-righting (with permission of [12]); (**4**) length–volume relationships of cylinders depend on the fiber angle α between the spiral netted fibers and the longitudinal axis of the cylinder; (**5**) curved 3D reticulated demonstrator with asymmetric deformation: connecting the net meshes on the adaxial side and/or cutting the net meshes on the abaxial side result in a large curvature of the tube when inflated; (**6**) biomimetic product to be developed: 3D reticulated actuator for use in technology.

2. Biology Push Process (Biomimetic Bottom-Up Approach)

2.1. Biological Question

The observation that tilted stems of papaya (*Carica papaya* L.; hereafter: *C. carica*) and leaning young trees of balsa (*Ochroma pyramidale* (Cav. ex Lam.) Urb.; hereafter: *O. pyramidale*) can reorient rapidly leads to the question of the underlying principles [11,27] (Figure 3). An understanding of the functional principle makes it all the more interesting, because structures similar to those of balsa and papaya have been found in *Lyginopteris oldhamia* (hereafter: *L. oldhamia*) and may provide information about the form–structure– function relationship in a fossil plant [12].

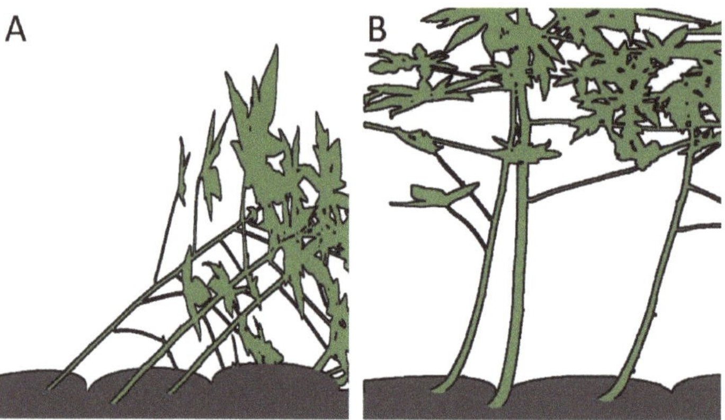

Figure 3. Erectile movement of *Carica papaya* stems. (**A**) Trees planted in tilted orientation with an inclination of 45°; (**B**) up-righting after 20 days resulting from active bending by growth of the stems mainly in the basal stem region. (**A,B**) redrawn and modified from [11].

2.2. Biological Concept Generators

C. carica and *O. pyramidale* have in common that their outer cortex can be unequally strained because of asymmetric growth of their inner secondary tissues. However, they do not use fiber structures at the cell wall level, but instead employ, at a higher hierarchical tissue level, netted fiber structures enveloping the whole stem (Figure 4A,B).

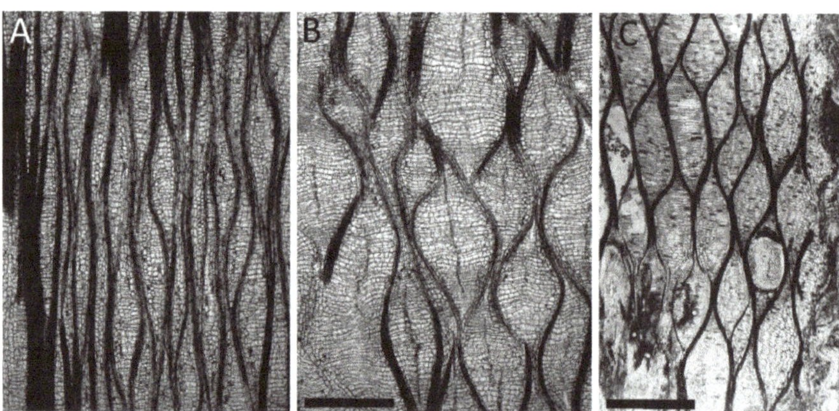

Figure 4. Tangential sections of the outer cortices of *Carica papaya* (**A,B**) and *Lyginopteris oldhamia* (**C**). (**A**) Unstrained abaxial side, (**B**) highly strained adaxial side, and (**C**) strained "Dictoxylon" cortex. Scale bars: (**A** = **B**) – 1 mm, (**C**) – 3 mm.

In fossil records of *L. oldhamia*, an extinct species common in the Upper Carboniferous of Europe (300 Myr B.C.), we can find similar structures in the so-called "Dictyoxylon" outer cortex. The "Dictyoxylon" is believed to be able to stretch following the stresses imposed by the increase in volume of the inner secondary tissues [28,29]. Similar to the cortex structure found in *C. carica* and *O. pyramidale*, the "Dictyoxylon" cortex of *L. oldhamia* is composed of sclerenchymatous strands that are imbedded in a parenchymatous ground tissue and that form a net-like structure (Figure 4). Moreover, [12] have shown that asymmetric secondary growth can also be present in the fossil stems of *L. oldhamia*, and that straining of the cortex is higher in the cortex above zones of increased secondary growth (Figure 4C).

2.3. Functional Principle

Because of the unequal straining that occurs in the outer cortex and that is specifically caused by asymmetric secondary growth, the described extant plants are able to straighten inclined (non-vertical) stems (Figure 3) [27]. The asymmetric increase in volume of the vascular secondary tissues leads to higher straining (and thereby shortening) of the adaxial (upper) side than the abaxial (lower) side, resulting in a curvature of the axis (Figure 5). This functional principle is highly dependent on the angles of the net-like cortex structures in *C. papaya* and *O. pyramidale*. Because of structural similarities, we can even assume that the fossil plant *L. oldhamia* developed a comparable functional principle and was also able to straighten its axis [12,30].

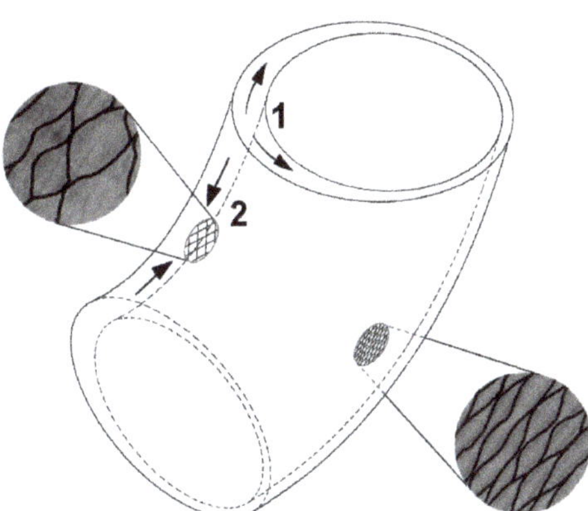

Figure 5. Erectile movement in *Carica papaya* and *Ochroma pyramidale* stems. Asymmetric growth in thickness of the secondary xylem creates stronger circumferential stresses (arrows 1) on the adaxial side of the stem (additional secondary wood formation); these stresses are partially converted into axial shortening (arrows 2) by the wrapping bark with its crossed net-like fiber bundles. Modified from [12].

2.4. Abstraction

To understand the functional principle of the biological concept generators described above, we first examined the simplified case of cylinders with braided fibers and equal straining before we proceeded to structures allowing for curvature.

Under internal pressure, the interaction of axial and (circumferential) tangential stresses causes braided cylinders with an angle α, between the longitudinal axis and the fibers, greater than 54.7° to lengthen and become thinner, while braided cylinders with an angle less than 54.7° shorten and thicken (Figure 6). An angle of 54.7° is the so-called neutral angle at which the volume under internal pressure remains constant (highest), and at which neither shortening nor elongation of the cylinder occurs [31].

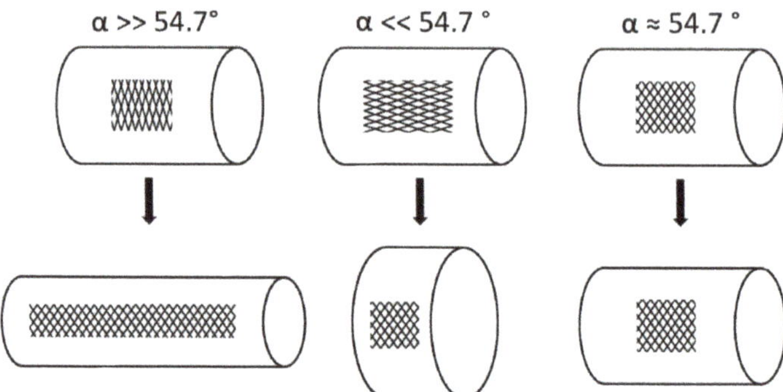

Figure 6. Length–volume relationships of cylinders surrounded by spiral netted fibers. The fiber angle α is the angle between the spiral netted fibers and the longitudinal axis of the cylinder. Maximal volume is reached when $\alpha = 54.7°$. When the initial α is greater than 54.7°, the cylinder elongates and becomes thinner under internal pressure, whereas if the initial α is less than 54.7°, the cylinder shortens and thickens. At $\alpha = 54.7°$, the length and volume remain constant under internal pressure.

Why should this angle be 54.7°? In a cylinder that is pressurized from the inside and that is in a state of equilibrium, the circumferential stresses are always twice as high as the axial stresses [32], i.e., we can determine a theoretical angle α' by abstracting the mesh in the cylinder, in a first step, to a two-dimensional right-angled triangle (=simplified force parallel diagram) (Figure 7A) with an edge length $2a$ (corresponding to the assumed vector of the circumferential stress) and a vertical edge length a (corresponding to the assumed vector of the axial stress) (Figure 7B). For this simplified two-dimensional mesh with twice as much stress in one direction as in another (perpendicular) direction, we obtain an angle of $\alpha' = \tan^{-1}(2) = 63.43°$.

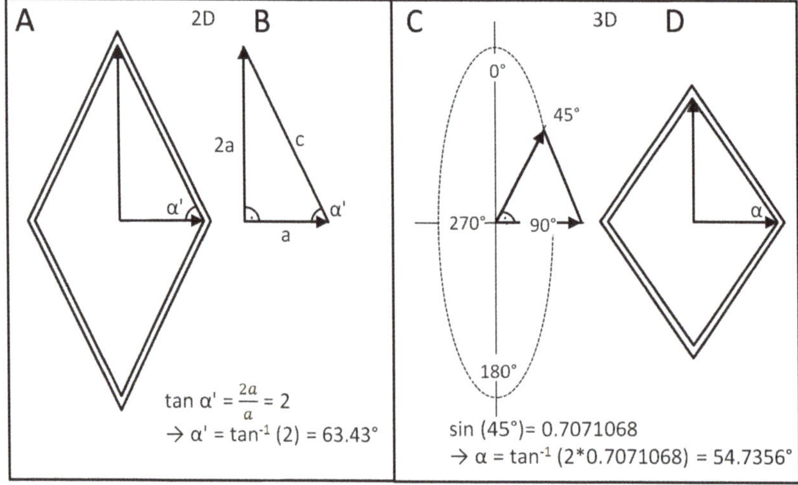

Figure 7. Geometric representations for determination of the neutral angle. (**A,B**) The angle α' can be defined in a 2D case by the vector a (parallel to the longitudinal axis of the cylinder) and a vector $2a$ (vertical to the longitudinal axis of the cylinder) for a simplified two-dimensional case. (**C,D**) Taking into account the rotational axis, α' can be transformed into the neutral angle α.

However, in contrast to this simplified 2D model, the cylinder is a three-dimensional object, i.e., the simplified "$2a/a$" triangle is tilted on average by 45° to the two-dimensional plane. The value of 45° results as the mean value between the two extreme cases: (1) parallel to a 2D plane (i.e., tilted by 0 degrees) and (2) perpendicular to it (i.e., tilted by 90°). Thus, the real (3D) vector of circumferential stress is not $2a$ but $2a \times \sin 45° = 2a \times 0.707 = 1.414$, since the mesh is not in a flat plane but in a plane of rotation around the central longitudinal axis of the cylinder (Figure 7C). Therefore, the real neutral angle $\alpha = \tan^{-1} (1.41a/a) = \tan^{-1} (1.414) = 54.7°$ (Figure 7D).

This neutral angle is important, because it is used in many technical tubular structures in which radial expansion and axial contraction take place, such as pneumatic muscles reported by Festo [21] (Figure 2A), and other fluidic muscles [33]. The flexibility of the structure is also used in the so-called stents, which can be inserted into large blood vessels in order to avoid blockage. These stents are usually braided at an angle equal to or close to 54.7°. Equally important is the production of other tubes with mesh structures that are arranged at a neutral angle, e.g., in water-conducting hoses (Figure 2B). Here, shortening or expansion is undesirable; the aim is to enable the highest possible internal pressures without significant longitudinal/transverse deformation of the hose and to maintain volume.

In *C. papaya*, *O. pyramidale*, and *L. oldhamia*, the angle α between the fibrous cortex structures and the longitudinal axis is significantly smaller than 54.7° (see Figure 4) so that a "pressurization," i.e., stresses generated by the increase in volume of secondary tissues, leads to the shortening of the structure. Because of the asymmetry of the vascular secondary growth, the structure is only shortened on one side, and thereby, a curvature of the axes can be evoked and modified by the amount of asymmetric secondary growth.

2.5. Feasibility Studies

In this article, we will use the manufacturing of simple functional demonstrators to demonstrate the way in which insights into functional principles from living nature can be transferred into technical implementations. We will first reproduce the functional principle of linear shortening or elongation and then build a further developed (biomimetic) demonstrator with which a directional adjustment, i.e., bending and curving, can be achieved, as in the concept generators *O. pyramidale* and *C. papaya*.

2.5.1. Technical Translation as Biomimetic Demonstrator with Linear Deformation

We can demonstrate the deformation of the cross-linked netted tube structure in a simple experiment based on the experiments first published in [22]. For this simple proof of concept, we use inflatable elements (balloons) and net-like tube structures (so-called bottle nets). Inflation of the balloons in the bottle nets leads to a radial expansion and an axial shortening of the structure: we measured axial shortening to 90% of the original length and radial expansion to 125% of the original width, as the initial angle of the net fibers is 15° (i.e., much smaller than 54.7°) (Figure 8A,B). If, on the other hand, we cut open the bottle nets and connect them in the other direction by using cable ties, the angle is 90–15 = 75° and, theoretically, a radial thinning and axial expansion is expected. The measured axial elongation amounts to 30%, i.e., the structure elongates under internal pressure to 130% of its original length. However, no radial thinning is measurable (Figure 8C,D). The reasons that our simple experiment (proof of concept) does not work exactly as predicted by theory are as follows: (1) the balloons do not expand ideally in a barrel shape (but in a more spherical shape), despite the fibers being tied around them, (2) the nets do not form ideal rhomboid meshes but assume other patterns when they are deformed. We therefore do not find the postulated 54.7° mesh fiber angle but an approximation to it (compared with the angle of the initial structure) and (3) the net and the balloon are not firmly connected to each other at each end.

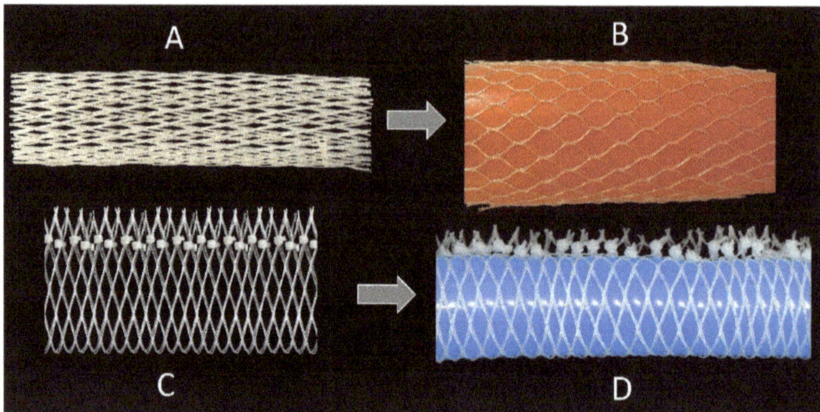

Figure 8. Linear mesh demonstrator. Change in the mesh geometry of bottle nets with an original angle considerably smaller than 54.7° (**A**), and considerably larger than 54.7° (**C**) and an additional fixation by cable ties that allows for lengthening. Inflation of the balloon leads to (**B**) a radial expansion and an axial shortening and (**D**) an axial elongation but no radial thinning.

2.5.2. Technical Translation as Biomimetic Demonstrator with Asymmetric Deformation

We created a special type of demonstrator by integrating a structural asymmetry into the tubular structure, following the example of the plants *O. pyramidale* and *C. papaya*. In the species mentioned, secondary thickness growth is asymmetric, causing a directional difference in internal pressure (see Section 2.2), as the structures expand more in the areas of increased tension caused by the increase in volume of the secondary tissues (see the description of the process above). Such a "directed" internal pressure is difficult to achieve with air-filled structures. Therefore, to produce a demonstrator with asymmetric expansion, the surrounding net-like structures have to be modified. We can adapt the bottle nets by obtaining larger meshes by cutting each second mesh on the adaxial side (here, rhombs are cut out as demonstrated in Figure 9A) so that they can expand more easily on the adaxial side than on the opposite side with the original denser meshing (the abaxial side; here, the meshes are additionally connected to each other by a household string to prevent expansion). A considerable curvature of the structure now occurs during inflation of the balloon (Figure 9B,C). The iterative functional principle of the technical demonstrator works in an "opposite" way to the natural models: initially uniform radial expansion results in a curvature as fewer fiber meshes, i.e., less opposing resistance, are present on one side "from the outset," i.e., the asymmetrically arranged fibers cause the curvature. This is in contrast to the natural role models, in which the asymmetry of the fibers is a result of the directional secondary growth, i.e., an asymmetric increase in inner volume. Nevertheless, the result is similar: a curvature can be created. This simple example shows that, in bio-inspired implementation, both an understanding and an abstraction of the functional principle are important and are not the consequence of a 1:1 imitation.

Figure 9. Curved mesh demonstrator. (**A**) Reduction of the resistance on the abaxial side by cutting out one rhombus in each repeat. (**B**) Connection of the net meshes on the adaxial side together with cutting each second mesh on the abaxial side results in (**C**) a pronounced curvature of the tube when inflated.

3. Discussion

A comparison of the biological model structures and the bio-inspired technical demonstrators shows similarities and dissimilarities. Similarities arise from the abstraction of the functional principle and its translation into a common language for natural scientists and engineers. Differences arise because we do not use the biological model as a blueprint but, instead, have to reinterpret the functional principle for use in the technical field and produce the biomimetic product from technical materials. Moreover, we can even go beyond a single biological model and combine the functional principles of several models in the biomimetic product.

Similarities between the biological models and the technical application are as follows:

- *Structure:* The structural setup, i.e., a tubular geometry surrounded by a peripheral network of fibers, is similar in *C. papaya*, *O. pyramidale*, *L. oldhamia*, and the demonstrators of the bio-inspired 3D reticulated actuator.
- *Change of shape:* Length–volume relationship depends on the fiber angle α between the spiral netted fibers and the longitudinal axis of the cylinder: $\alpha = 54.7°$ leads to a maximal volume, but, under internal pressure, the cylinder elongates and thins when $\alpha \gg 54.7°$ and shortens and thickens when $\alpha \ll 54.7°$.

Dissimilarities between the biological models and the technical application are as follows:

- *Actuation:* Shortening/elongation of the cylindrical net structures is often hydraulically driven in nature, e.g., the swelling of the so-called G-layer [34] in cell walls of reaction (tension) wood on the upper side of branches in deciduous trees. The increase in volume of the cylindrical cellular structure that occurs here is converted into a shortening of the wood cells in the axial longitudinal direction by a special enveloping layer [5,6]. Thus, tension is generated in the tension wood of deciduous trees. An additional example of this actuation principle by liquids is the movement of the feet of sea urchins and starfish, in which extension or contraction is actuated by pressure changes in the ambulacral system [13]. On a different scale, the similar deformations of the netted fibers in *C. papaya*, *O. pyramidale*, and *L. oldhamia* are actuated by growth-induced stresses. All of these examples have in common that the actuation is markedly different from the pneumatic pressurization used in the demonstrators (main difference: incompressible versus compressible actuation agents, see below).
- *Speed of movement:* Re-orientation of plant stems takes place by slow growth-induced processes, both in tension wood and in cortex-based (re-)erection, compared with the rapid inflation-based size and shape change in the demonstrator. Nevertheless, the time frame of the movements achieved with the bio-inspired demonstrators above shows more similarities with that of the shape changes in the feet of sea urchins and starfish.

- *Reversibility:* Re-orientation in the plants is irrevocable, whereas the pressurization of the demonstrators can easily be undone; this again shows similarities with the reversible motion of the feet of sea urchins and starfish.
- *Power:* Forces exerted by liquids (turgor pressure, swelling of G-layers, liquid-filled feet of sea urchins) or solids (secondary tissues) are higher than the forces attainable by inflation because of the incompressibility of liquids and solids as compared with compressible air.
- *Curvature actuation:* As detailed above, curvature is the result of asymmetric growth in *C. papaya*, *O. pyramidale*, and *L. oldhamia*, whereas it is a consequence of the asymmetric net structure in the technical translation.

4. Conclusions and Outlook

Technical structures based on the functional principle described above might find application in the automotive industry and in aerospace, i.e., areas in which lightweight construction and adaptive shape adjustments are of particular interest. Such implementation is not necessarily limited to pneumatic structures: in the field of architecture, similar adaptive hulls can help to improve the mechanical properties of concrete sheathing by using FRP (fiber-reinforced composites). Here, the combination of bio-inspired braided structures [35–38] and the cross-linking presented in this study has a high potential for successful transfer. In addition, the above actuation can be developed and improved further by replacing the internal pressurized cylinder by reactive materials embedded in the meshes of the net-like hull, which isotropically or anisotropically swell or shrink upon external stimuli such as changes in humidity, pH, temperature, or light conditions. This would leave the entire inner volume of the 3D reticulated actuator free for additional integrated functions. In another implementation, the braided external structure could allow for various advantages in the spaces between the braided filaments, enabling transfer of light, gasses, and fluids in the absence of a continuous layer.

The described simple, easy-to-build, and inexpensive demonstrators not only can serve as a basis for technical implementations, but might also be well suited for teaching at university level and for education in high schools.

Author Contributions: Conceptualization, T.M. and O.S.; methodology, T.M. and T.S.; writing—original draft preparation, T.M. and O.S.; writing—review and editing, T.M., O.S., and T.S.; funding acquisition, T.S. and O.S. All authors have read and agreed to the published version of the manuscript.

Funding: This research was funded by the Deutsche Forschungsgemeinschaft (DFG, German Research Foundation) under Germany's Excellence Strategy—EXC-2193/1–390951807.

Institutional Review Board Statement: Not applicable.

Informed Consent Statement: Not applicable.

Data Availability Statement: Data is contained within the article.

Acknowledgments: We thank R. Theresa Jones for improving the English.

Conflicts of Interest: The authors declare no conflict of interest. The funders had no role in the design of the study; in the collection, analyses, or interpretation of data; in the writing of the manuscript, or in the decision to publish the results.

References

1. Poppinga, S.; Joyeux, M. Different mechanics of snap-trapping in the two closely related carnivorous plants *Dionaea muscipula* and *Aldrovanda vesiculosa*. *Phys. Rev.* **2011**, *84*, 041928. [CrossRef]
2. Poppinga, S.; Bauer, U.; Speck, T.; Volkov, A.G. Motile traps. In *Carnivorous Plants: Physiology, Ecology, and Evolution*; Ellison, A.M., Adamec, L., Eds.; Oxford University Press: Oxford, UK, 2017; pp. 180–193. [CrossRef]
3. Poppinga, S.; Daber, L.E.; Westermeier, A.S.; Kruppert, S.; Horstmann, M.; Tollrian, R.; Speck, T. Biomechanical analysis of prey capture in the carnivorous Southern bladderwort (*Utricularia australis*). *Sci. Rep.* **2017**, *7*, 1776. [CrossRef] [PubMed]
4. Westermeier, A.S.; Fleischmann, A.; Müller, K.; Schäferhoff, B.; Rubach, C.; Speck, T.; Poppinga, S. Trap diversity and character evolution in carnivorous bladderworts (*Utricularia*, Lentibulariaceae). *Sci. Rep.* **2017**, *7*, 12052. [CrossRef] [PubMed]

5. Goswami, L.; Dunlop, J.W.; Jungnikl, K.; Eder, M.; Gierlinger, N.; Coutand, C.; Jeronimidis, G.; Fratzl, P.; Burgert, I. Stress generation in the tension wood of poplar is based on the lateral swelling power of the G-layer. *Plant J.* **2008**, *56*, 531–538. [CrossRef]

6. Burgert, I.; Fratzl, P. Actuation systems in plants as prototypes for bioinspired devices. *Philos. Trans. R. Soc.* **2009**, *367*, 1541–1557. [CrossRef]

7. Speck, T.; Poppinga, S.; Speck, O.; Tauber, F. Bio-inspired life-like motile materials systems: Changing the boundaries between living and technical systems in the Anthropocene? *Anthr. Rev.* **2021**. under review.

8. Skotheim, J.M.; Mahadevan, L. Physical limits and design principles for plant and fungal movements. *Science* **2005**, *308*, 1308–1310. [CrossRef] [PubMed]

9. Dumais, J.; Forterre, Y. "Vegetable dynamicks": The role of water in plant movements. *Annu. Rev. Fluid Mech.* **2012**, *44*, 453–478. [CrossRef]

10. Correa, D.; Poppinga, S.; Mylo, M.; Westermeier, A.S.; Bruchmann, B.; Menges, A.; Speck, T. 4D pine scale: Biomimetic 4D printed autonomous scale and flap structures capable of multi-phase movement. *Philos. Trans. R. Soc.* **2020**, *378*, 20190445. [CrossRef]

11. Kempe, A.; Lautenschläger, T.; Neinhuis, C. Reorientation in tilted stems of papaya by differential growth. *Int. J. Plant Sci.* **2014**, *175*, 537–543. [CrossRef]

12. Masselter, T.; Kempe, A.; Caliaro, S.; Neinhuis, C.; Speck, T. Comparing structure and biomechanics of extant *Carica papaya* and *Ochroma pyramidale* stems allows re-evaluating the functional morphology of the fossil 'seed fern' *Lyginopteris oldhamia. Rev. Palaeobot. Palynol.* **2017**, *246*, 258–263. [CrossRef]

13. McCurley, R.S.; Kier, W.M. The functional morphology of starfish tube feet: The role of a crossed-fiber helical array in movement. *Biol. Bull.* **1995**, *188*, 197–209. [CrossRef] [PubMed]

14. Heydari, S.; Johnson, A.; Ellers, O.; McHenry, M.J.; Kanso, E. Sea star inspired crawling and bouncing. *J. R. Soc. Interface* **2020**, *17*, 20190700. [CrossRef] [PubMed]

15. Jonas, F.A.; Born, L.; Möhl, C.; Hesse, L.; Bunk, K.; Masselter, T.; Speck, T.; Gresser, G.; Knippers, J. *Biomimetics for Architecture: Learning from Nature*; Knippers, J., Schmid, U., Speck, T., Eds.; Birkhäuser Verlag: Basel, Switzerland, 2019.

16. Poppinga, S.; Zollfrank, C.; Prucker, O.; Rühe, J.; Menges, A.; Cheng, T.; Speck, T. Towards a new generation of smart biomimetic actuators for architecture. *Adv. Mater.* **2018**, *30*, 1703653. [CrossRef] [PubMed]

17. Correa, D.; Krieg, O.D.; Menges, A.; Reichert, S.; Rinderspacher, K. HygroSkin: A climate-responsive prototype project based on the elastic and hygroscopic properties of wood. *Adapt. Archit.* **2013**, 33–41.

18. Menges, A.; Knippers, J. *Architecture Research Building ICD/ITKE, 2010–2020*; Birkhäuser Verlag: Basel, Switzerland, 2020.

19. Mader, A.; Langer, M.; Knippers, J.; Speck, O. Learning from plant movements triggered by bulliform cells: The biomimetic cellular actuator. *J. R. Soc. Interface* **2020**, *17*, 20200358. [CrossRef]

20. Körner, A.; Born, L.; Mader, A.; Sachse, R.; Saffarian, S.; Westermeier, A.S.; Poppinga, S.; Bischoff, M.; Gresser, G.T.; Milwich, M.; et al. Flectofold—a biomimetic compliant shading device for complex free form facades. *Smart Mater. Struct.* **2017**, *27*, 017001. [CrossRef]

21. Festo SE & Co. KG. Fluidic Muscle DMSP/MAS. Operating Instructions. 2018. p. 18. Available online: https://www.festo.com/net/SupportPortal/Files/492257/DMSP_MAS_2018-01f_8081496g1.pdf (accessed on 1 April 2021).

22. Speck, O.; Boblan, I. Vom Luftballon zum künstlichen Muskel. *Grundsch. Sachunterr.* **2014**, *62*, 20–26.

23. Behrens, R.; Poggendorf, M.; Schulenburg, E.; Elkmann, N. An elephant's trunk-inspired robotic arm—Trajectory determination and control. In Proceedings of the ROBOTIK 2012, 7th German Conference on Robotics, Munich, Germany, 21–22 May 2012; pp. 1–5, ISBN 978-1-5231-0924-1.

24. Boblan, I.; Schulz, A.; Tuchscherer, A.; Perfilov, I.; Bertrand, B. A compliant lightweight universal joint cascadable to a multi-joint kinematics—Tripedale Alternanzkaskade TAK. In Proceedings of the AMAM 2013, 6th International Symposium zur Bewegungsforschung, Darmstadt, Germany, 11–14 March 2013; p. 4. Available online: http://www.biorobotiklabor.de/pdfs/boblan_2013_AMAM2013.pdf (accessed on 22 March 2021).

25. Speck, T.; Speck, O. Process sequences in biomimetic research. In *Design and Nature IV*; Brebbia, C.A., Ed.; WIT Press: Southampton, UK, 2008; pp. 3–11. ISBN 978-1-84564-120-7.

26. ISO 18458:2015-05. *Biomimetics—Terminology, Concepts and Methodology*; Beuth: Berlin, Germany, 2015.

27. Fisher, J.B.; Müller, R.J. Reaction anatomy and reorientation in leaning stems of balsa (*Ochroma*) and papaya (*Carica*). *Can. J. Bot.* **1983**, *61*, 880–887. [CrossRef]

28. Masselter, T.; Speck, T.; Rowe, N.P. Ontogenetic reconstruction of the Carboniferous seed plant *Lyginopteris oldhamia. Int. J. Plant Sci.* **2006**, *167*, 147–166. [CrossRef]

29. Masselter, T.; Speck, T. Secondary growth stresses in recent and fossil plants: Physical and mathematical modeling. *Rev. Paleobot. Palynol.* **2014**, *201*, 47–55. [CrossRef]

30. Poppinga, S.; Speck, T. Bark, the neglected tree postural motor system. *New Phytol.* **2019**, *221*, 7–9. [CrossRef] [PubMed]

31. Horgan, C.O.; Murphy, J.G. Magic angles in the mechanics of fibrous soft materials. *Mech. Soft Mater.* **2019**, *1*, 2. [CrossRef]

32. Gordon, J.E. *The New Science of Strong Materials, or Why You Don't Fall through the Floor*; Princeton University Press: Princeton, NJ, USA, 1976; p. 287.

33. Takosoglu, J.E.; Laski, P.A.; Blasiak, S.; Bracha, G.; Pietrala, D. Determining the static characteristics of pneumatic muscles. *Meas. Control* **2016**, *49*, 62–71. [CrossRef]

34. Gorshkova, T.; Chernova, T.; Mokshina, N.; Ageeva, M.; Mikshina, P. Plant 'muscles': Fibers with a tertiary cell wall. *New Phytol.* **2018**, *218*, 66–72. [CrossRef] [PubMed]
35. Born, L.; Jonas, F.A.; Bunk, K.; Masselter, T.; Speck, T.; Knippers, J.; Gresser, G.T. Branched structures in plants and architecture. In *Biomimetic Research for Architecture and Building Construction—Biological Design and Integrative Structures*; Knippers, J., Speck, T., Nickel, K., Eds.; Springer International Publishing: Cham, Switzerland, 2016; pp. 195–215, ISBN 978-3-319-46374-2.
36. Bunk, K.; Jonas, F.A.; Born, L.; Hesse, L.; Möhl, C.; Gresser, G.; Knippers, J.; Speck, T.; Masselter, T. From plant branchings to technical support structures. In *Biomimetics for Architecture: Learning from Nature*; Knippers, J., Schmid, U., Speck, T., Eds.; Birkhäuser: Basel, Switzerland, 2019; pp. 144–152, ISBN 987-3-0356-1786-3.
37. Hesse, L.; Leupold, J.; Poppinga, S.; Wick, M.; Strobel, K.; Speck, T.; Masselter, T. Resolving form-structure-function relationships in plants with MRI for biomimetic transfer. *Integr. Comp. Biol.* **2019**, *59*, 1713–1726. [CrossRef] [PubMed]
38. Jonas, F.A.; Born, L.; Möhl, C.; Hesse, L.; Bunk, K.; Masselter, T.; Speck, T.; Gresser, G.; Knippers, J. New branched loadbearing structures in architecture. In *Biomimetics for Architecture: Learning from Nature*; Knippers, J., Schmid, U., Speck, T., Eds.; Birkhäuser: Basel, Switzerland, 2019; pp. 153–162, ISBN 987-3-0356-1786-3.

Article

Strengthening Structures in the Petiole–Lamina Junction of Peltate Leaves

Julian Wunnenberg, Annabell Rjosk *, Christoph Neinhuis and Thea Lautenschläger

Department of Biology, Faculty of Science, Institute of Botany, Technische Universität Dresden, 01062 Dresden, Germany; julian.wunnenberg@tu-dresden.de (J.W.); christoph.neinhuis@tu-dresden.de (C.N.); thea.lautenschlaeger@tu-dresden.de (T.L.)
* Correspondence: annabell.rjosk@tu-dresden.de

Abstract: Peltate- or umbrella- shaped leaves are characterised by a petiole more or less centrally attached to the lamina on the abaxial side. The transition from the petiole to lamina in peltate leaves resembles a significant and abrupt geometrical change from a beam to a plate in a very compact shape. Since these leaves have not been subject of many studies, the distribution of that specific leaf morphology in the plant kingdom was investigated. Furthermore, the connection between the petiole and lamina of several peltate species was studied anatomically and morphologically, focusing on the reinforcing fibre strands. We found peltate leaves in 357 species representing 25 orders, 40 families and 99 genera. The majority are herbaceous perennials growing in shady, humid to wet habitats mainly distributed in the subtropical–tropical zones. Detailed anatomical investigation of 41 species revealed several distinct principles of how the transition zone between the petiole and lamina is organised. In-depth analysis of these different types accompanied by finite element-modelling could serve as inspiration for supporting structures in lightweight construction.

Keywords: peltate leaves; petiole; petiole–lamina junction; anatomy; strengthening structures

Citation: Wunnenberg, J.; Rjosk, A.; Neinhuis, C.; Lautenschläger, T. Strengthening Structures in the Petiole–Lamina Junction of Peltate Leaves. *Biomimetics* **2021**, *6*, 25. https://doi.org/10.3390/biomimetics6020025

Academic Editor: Josep Samitier

Received: 8 February 2021
Accepted: 31 March 2021
Published: 2 April 2021

Publisher's Note: MDPI stays neutral with regard to jurisdictional claims in published maps and institutional affiliations.

1. Introduction

Generally, leaves are important organs for plants, especially for photosynthesis and the production of organic compounds [1–4]. As such, the orientation of the leaves towards the sun is essential. Leaves are exposed to a multitude of mechanical stresses, caused by natural gravity and environmental factors such as wind and rain [1,5–7]. The petioles in particular must fulfil two important functions: (1) sustain their own weight and that of the lamina against gravity and (2) provide enough mechanical stability to withstand bending and twisting while being flexible enough not to be damaged [5,6,8,9]. Therefore, petioles can be described mechanically as elastic cantilevered beams [6,9,10].

To achieve the maximum rate of photosynthesis, the parameters of the leaf (shape, area, weight, petiole parameters, etc.) should be optimal [1]. However, there are many different leaf shapes in nature, each of which have different effects on the petioles [1,11]. The internal organisation of the petioles, including ground tissue and support or strengthening tissue and their respective proportions, are key parameters in mastering such mechanical requirements [1,9,11–13]. Among the tissues, parenchyma is the least differentiated with the lowest stiffness [1,9]. The highly variable collenchyma, also part of the ground tissue, is an important reinforcing structural element [1,12,14,15]. As both are viscoelastic tissues, stiffness and the contribution as strengthening tissues strongly depend on turgor pressure [1,9]. Besides the ground tissues, the petioles contain vascular tissue, formed by the phloem and xylem which serve as a pathway for water, nutrients and assimilates, both to and from the leaf blade [1,16]. The xylem also acts as a primary strengthening structure. Its stiffness is comparable to that of the sclerenchyma, which is the stiffest tissue in the petioles and often forms caps above the phloem or sclerenchymatic rings surrounding the vascular bundles [9,17].

Most leaves consist of a petiole and the leaf blade, but the peltate leaf shape is not very common and has not been the subject of many studies [18]. In 1932, Wilhelm Troll provided one of the first explanations for the peltate leaf shape and defined them by their petiole insertion point [19]. The petioles are attached to the lamina on the abaxial side either in the centre or approximated to the leaf margin, distinguishing peltate leaves from the common marginal petiole insertion [19]. The unifacial petioles have a ring-shaped arrangement of the vascular bundles [19–22]. Although they are relatively small in number, the appearance of peltate leaves can be very diverse, including entire and centro-peltate leaves (e.g., *Nelumbo nucifera* or *Victoria cruziana*), as well as species with palmatifid (e.g., *Darmera peltata*), lobed (e.g., *Podophyllum peltatum*) and palmately compound leaves (e.g., *Schefflera arboricola*) (Figure 1) [18,20].

Figure 1. Variety of peltate leaf shapes (according to Troll, 1932 and Ebel, 1998 [18,20]). (**A**)—orbicular entire (*Euryale ferox*), (**B**)—ovate entire with pointed apex (*Macaranga tanarius*), (**C**)—palmately compound with leaflets (*Schefflera arboricola*), (**D**)—palmatipartite (*Cecropia peltata*), (**E**)—palmatifid (*Darmera peltata*), (**F**)—orbicular crenate toothed (*Nymphaea lotus*). (**A,E**)—central petiole insertion, (**B–D,F**)—eccentric insertion of petiole.

The reason for the evolution of peltate leaves is not yet fully understood. First theories suggested the facilitated orientation towards the sun and thus, the efficient light absorption [19,23]. Friedrich Ebel (1998) introduced a connection between peltation and certain chorological and habitat conditions [18]. In fact, herbaceous plants with peltate leaves are predominantly geophytes with rhizomes, stolons, bulbs, or tubers and almost all species grow in wet, humid or alternately wet locations preferentially close to lakes, ponds, swamps, meadows, streams, on river banks, in forests or amongst humid, shady to semi-shady rocks [18]. While genera with peltately leaved species are more or less evenly distributed in subtropical, meridional and temperate areas with ± oceanic climate (especially in eastern North and Central America and South East Asia), they are less common, relative to the total number of genera, in boreal zones, steppes, semi-deserts and desert habitats [18]. Only a few sources are available dealing with the distribution of species with peltate leaves. The first attempt to compile such a list was made by C. de Candolle in 1899 [24,25]. After several updates, Ebel (1989) was the last to provide a comprehensive list of peltate taxa distributed in 54 out of currently 416 angiosperm families (without distinguishing between the peltate leaf shape or variants and taxa with ascidiate leaves, e.g., Nepenthaceae, Cephalotaceae, Sarraceniaceae, or the Droseraceae) [18,26].

While some studies are available on the morphology and anatomy of peltate leaves and their petioles [19–22], the transition zone from the petiole to lamina has been analysed so far only in a single study by Sacher et al. (2019) [13]. Focusing on two peltate plant species (one from Araceae and one from Tropaeolaceae), the study identified two different variants of strengthening structures present in the compact transition zone [13]. These structures represent two types of branching of the vascular strands, which have been shown to act as a stabilising element. Such biological strengthening structures may have great potential as the inspiration for applications in lightweight construction [13].

This study intends to provide an overview about the occurrence and distribution of the peltate leaf shape among flowering plants and to analyse the morphology and anatomy of the petiole and the petiole–lamina transition zone of different peltate species in a wider approach focusing on characterising and categorising strengthening structures in the transition zone. With two types of strengthening structures found by Sacher et al. (2019) [13], our research aims to determine if there are more distinct principles of how the petiole–lamina junction in peltate leaves can be organised.

2. Materials and Methods

2.1. Screening

In search of plant species with peltate leaves, the living collection of the Botanical Garden of Technische Universität Dresden and the literature about the peltate leaf shape such as Uittien (1929), Troll (1932, 1939, 1955), Roth (1949, 1952) or Franck (1976) served as basis [19–22,25,27]. Internet sources as well as online libraries complemented the search for further species.

In this study, the term "peltate leaf" refers exclusively to the morphology of the visible foliage. It focused only on the vegetative, but not floral parts of the plant, such as stamens or carpels (Troll 1992) or peltate petals (Leinfellner 1958) [19,28].

The present work focused on species with an eccentric (margin approximated) or centrally attached petiole, with lobed, crenate or pinnate peltate leaves. All taxa have been compiled in a list down to species level together with information about their origin and distribution as well as their habit, habitat and petiole insertion.

The species for which fresh material was available were examined microscopically. Information about the habit, habitat and ecology of additional species were collected from different databases:

- eFloras, Missouri Botanical Garden and Harvard University Herbaria [29];
- POWO (Plants of the World Online), Royal Botanic Gardens, Kew [30];
- GBIF (Global Biodiversity Information Facility), GBIF Secretariat Copenhagen [31].

The nomenclature is based on information from Tropicos® and © The Plant List (2013) [32,33]. The taxonomic classification follows the APG (Angiosperm Phylogeny Group) IV system of the Angiosperm phylogeny poster from 2019 [34]. The differentiation of peltation between centrally inserted and eccentrically inserted petioles is based on the definition in the "Manual of Leaf Architecture" by Ellis et al. (1999) [35]. The classification of the species distribution follows the list of the floral zones of the earth by Meusel et al. (1965–1992), edited and published by Jäger (2017) [36] and the floristic kingdoms by Pott (2014) [37].

2.2. Plant Material and Sampling

Plant samples were taken from the living collection of the Botanical Garden of the Technische Universität Dresden, Germany between September and December of 2019 and the Ecological–Botanical Garden of Universität Bayreuth, Germany in December 2019. All species were cultivated in the tropical greenhouses or under open-air conditions. Only non-damaged intact leaves and petioles were investigated (e.g., no frost or feeding damage) and a selection of small to large leaves. The leaf and the complete petiole were taken from the plant. To keep the samples as fresh as possible for direct analysis, they were transported to the laboratory in airtight containers and processed within a few hours. Samples of each species were preserved in 70% ethanol for later anatomical analysis.

2.3. Anatomy and Morphology

All fresh samples were scanned (Ricoh MP C3004ex, Ricoh Company, Ltd., Chuo, Tokyo, Japan) from both sides of the leaf using the program ImageJ (National Institutes of Health, Bethesda, MD, USA) to record the dimensions of leaves and petioles, including the length and width of the leaf and diameter of the petiole. In lobed or fully peltate leaves (almost circular), the midrib was used as orientation to determine the length of the leaf while the width (at the widest point) was measured perpendicular to the midrib. The leaf venation was separated from the intercostal areas of the lamina up to the second level of branching with a razor blade. The petioles were separated from the lamina and for both fresh and dry weight were determined using a precision scale (Mettler Toledo XA205DU, Mettler Toledo, Columbus, OH, USA). The fresh samples were dried for 5 days at 65 °C in a drying cabinet (Heraeus T12, Heraeus Instruments, Hanau, Germany) before measuring the dry weight.

Cross and longitudinal sections of leaves and petioles were prepared using a razor blade, stained with Astra-blue/Safranin (Morphisto GmbH, Frankfurt am Main, Germany) and observed with the reflected light microscope (Olympus SZX16, Olympus Corporation, Tokyo, Japan). Sections were photographed with a high-resolution microscope camera (Olympus DP6, Olympus Corporation, Tokyo, Japan). The number and area of the vascular bundles, the size of lignified areas and present types of tissue were measured for a small number of samples via images from the microscope camera using the image-processing program ImageJ (National Institutes of Health, Bethesda, MD, USA).

3. Results

3.1. Distribution of Peltate Species in the Plant Kingdom

The extent to which leaves can be described as peltate depends on the definition of the author. Therefore, the following list does not claim to be complete. Furthermore, several families or genera, with a high number of species with peltate leaves are not fully studied, e.g., *Begonia*, i.e., the list does not include all peltate leaved species of these genera or families.

The peltate leaf shape is found in all clades of angiosperms. In total, 357 species representing 99 genera, 40 families, and 25 orders were found (Table 1, for full list: see Supplementary Materials), including one order from the ANA grade (Nympheales) (ANA: Amborellales, Nymphaeales, Austrobaileyales), two from the magnoliids (Laurales and Piperales) and one order from the monocots (Alismatales). The remaining 21 orders belong to the eudicots. A fern genus with peltate leaved species could also be identified (Salviniales, *Marsilea*). Some families show a noticeably high number of peltate leaved species and were not studied completely, e.g., Piperaceae (two genera), Araceae (10 genera), Fabaceae (one genus) or Begoniaceae (one genus).

The analysis of additionally recorded parameters reveals that peltate species are preferentially distributed in the subtropical and tropical areas of the Neotropics and the Paleotropics. In South East Asia, the highest number of peltate leaved species are found in the paleotropic boreo-subtropical region, in Africa in the tropical zone, and in the Neotropics throughout the subtropical and tropical area. The southern austral zone contains considerably less peltate leaved genera than in the northern hemisphere, occurring mostly in special areas, such as the Capensis (*Oxalis, Ranunculus, Senecio*), New Zealand (*Ranunculus*) and the Juan Fernández Islands (*Gunnera*). *Tropaeolum, Gunnera* and *Ranunculus* are the only genera which are common in the Antarctic floristic kingdom. In the Holarctic, the main distribution areas are China (>15 genera) and meridional North America (6). The distribution of some genera also extends into the temperate and Boreal floral zone, while no representatives are recorded from the Arctic zone. The peltate leaved species of Europe are mainly found in the Mediterranean region in genera such as *Umbilicus, Hydrocotyle* or *Marsilea*.

Table 1. List of selected peltate species among vascular plants.

Clade	Order	Family	Species	Continent	Distribution	Habit	Habitat		Floral Zone	Floristic Kingdom	Peltation
F	Salviniales	Marsileaceae	*Marsilea ancylopoda* A. Braun	North America, South America	Tropical America, Subtropical America	h	p	a	boreostrop, trop, austrostrop, austr	Neo	c
F	Salviniales	Marsileaceae	*Marsilea batardae* Launert	Europe	Portugal, Spain	h	p	a	sm, m	Hol	c
F	Salviniales	Marsileaceae	*Marsilea macrocarpa* C. Presl	Africa	C Africa, S Africa, Madagascar	h	p	a	trop, austrostrop	Pal	c
F	Salviniales	Marsileaceae	*Marsilea oligospora* Goodd.	North America	USA	h	p	a	t, sm, m	Hol	c
F	Salviniales	Marsileaceae	*Marsilea strigosa* Willd.	Europe	Mediterranean region, Russia, Kazakhstan	h	-	a	sm	Hol	c
ANA	Nymphaeales	Cabombaceae	*Brasenia schreberi* J.F.Gmel.	Africa, South America, Asia	worldwide	h	p	a	austr, strop, trop, m, sm, temp	Hol, Pal, Neo, Aus	e
ANA	Nymphaeales	Cabombaceae	*Cabomba aquatica* Aubl.	South America	S America	h	p	a	strop, trop	Neo	c
ANA	Nymphaeales	Nymphaeaceae	*Euryale ferox* Salisb. ex K.D.Koenig and Sims	Asia	Asia	h	a,p	a	boreostrop, m	Pal	c
ANA	Nymphaeales	Nymphaeaceae	*Nymphaea colorata* Peter	Africa	Tanzania	h	p	a	trop	Pal	e
ANA	Nymphaeales	Nymphaeaceae	*Nymphaea gigantea* Hook.	Australia	Australia	h	p	a	austrostrop	Aus	e
ANA	Nymphaeales	Nymphaeaceae	*Nymphaea lotus* L.	Africa	Africa	h	p	a	strop, trop	Pal	e
ANA	Nymphaeales	Nymphaeaceae	*Victoria amazonica* (Poepp.) J.C. Sowerby	South America	S America	h	p	a	trop	Neo	c
ANA	Nymphaeales	Nymphaeaceae	*Victoria cruziana* Orb.	South America	S America	h	p	a	austrostrop	Neo	c
MAG	Laurales	Hernandiaceae	*Hernandia nymphaeifolia* (C.Presl) Kubitzki	Asia, Africa	SE Asia, Madagascar	l	p	t	trop	Pal	e
MAG	Laurales	Hernandiaceae	*Hernandia sonora* L.	North America	Mexico, Caribbean	l	p	t	boreostrop	Neo	e
MAG	Piperales	Piperaceae	*Peperomia argyreia* E.Morr.	South America	Brazil	h	p	t	austrostrop	Neo	e
MAG	Piperales	Piperaceae	*Peperomia bracteata* A.W.Hill	North America	Guatemala, Mexico	h	p	t	boreostrop	Neo	c
MAG	Piperales	Piperaceae	*Peperomia cyclaminoides* A.W. Hill	South America	Bolivia	h	-	t	boreostrop	Neo	c
MAG	Piperales	Piperaceae	*Peperomia monticola* Miq.	North America	Mexico	h	p	t	boreostrop	Neo	e
MAG	Piperales	Piperaceae	*Peperomia sodiroi* C.DC.	South America	Ecuador	h	p	t	trop	Neo	e
MAG	Piperales	Piperaceae	*Piper peltatum* L.	North America, South America	S America, C America	h,l	p	t	strop, trop	Neo	e
MAG	Piperales	Piperaceae	*Piper fragile* Benth.	Asia	C Malesia, W Pacific	-	-	t	trop	Pal, Aus	e
MAG	Piperales	Piperaceae	*Piper peltatifolium* C.Y. Hao, H.S. Wu, Y.H. Tan	Asia	China	l	p	t	boreostrop	Pal	e

Table 1. *Cont.*

Clade	Order	Family	Species	Continent	Distribution	Habit	Habitat	Floral Zone	Floristic Kingdom	Peltation
MON	Alismatales	Araceae	*Alocasia cuprea* K.Koch	Asia	Borneo	h p	t	trop	Pal	e
MON	Alismatales	Araceae	*Alocasia fallax* Schott	Asia	Bangladesh, India	h p	t	boreostrop	Pal	e
MON	Alismatales	Araceae	*Alocasia longiloba* Miq.	Asia	SE Asia	h p	t	boreostrop, trop	Pal	e
MON	Alismatales	Araceae	*Alocasia peltata* M. Hotta	Asia	Borneo	h p	t	trop	Pal	e
MON	Alismatales	Araceae	*Alocasia reversa* N.E. Br.	Asia	Borneo	h p	t	trop	Pal	e
MON	Alismatales	Araceae	*Amorphophallus konjac* K.Koch	Asia	China, SE Asia	h p	t	m	Hol	e
MON	Alismatales	Araceae	*Anthurium forgetii* N.E.Br.	South America	Colombia	h p	t	trop	Neo	e
MON	Alismatales	Araceae	*Anthurium jureianum* Cath. and Olaio	South America	Brazil	h p	t	austrostrop	Neo	e
MON	Alismatales	Araceae	*Ariopsis peltata* Nimmo	Asia	India	h p	t	boreostrop, trop	Pal	e
MON	Alismatales	Araceae	*Ariopsis proanthera* N.E. Br.	Asia	Bangladesh, India, Myanmar, Nepal, Thailand	h p	t	boreostrop, trop	Pal	e
MON	Alismatales	Araceae	*Arisaema caudatum* Engl.	Asia	India	h p	t	boreostrop, trop	Pal	c
MON	Alismatales	Araceae	*Arisaema ciliatum* H.Li	Asia	China	h p	t	m	Hol	c
MON	Alismatales	Araceae	*Arisaema fischeri* Manudev and Nampy	Asia	India	h p	t	boreostrop, trop	Pal	c
MON	Alismatales	Araceae	*Arisaema peltatum* C.E.C. Fisch.	Asia	India	h p	t	boreostrop, trop	Pal	e
MON	Alismatales	Araceae	*Arisaema subulatum* Manudev and Nampy	Asia	India	h p	t	boreostrop, trop	Pal	c
MON	Alismatales	Araceae	*Caladium bicolor* Vent.	North America, South America	C America, S America	h p	t	boreostrop, trop	Neo	e
MON	Alismatales	Araceae	*Caladium clavatum* Hett., Bogner and J. Boos	South America	Ecuador	h p	t	trop	Neo	e
MON	Alismatales	Araceae	*Caladium humboldtii* (Raf.) Schott	South America	Brazil, Venezuela	h p	t	trop	Neo	e
MON	Alismatales	Araceae	*Caladium smaragdinum* K. Koch and C.D. Bouché	South America	Colombia, Venezuela	h p	t	trop	Neo	e
MON	Alismatales	Araceae	*Caladium steudnerifolium* Engl.	South America	Bolivia, Colombia, Ecuador, Peru	h p	t	trop, austrostrop	Neo	e
MON	Alismatales	Araceae	*Colocasia boyceana* Gogoi and Borah	Asia	India	h p	t	boreostrop, trop	Pal	e

Table 1. *Cont.*

Clade	Order	Family	Species	Continent	Distribution	Habit	Habitat	Floral Zone	Floristic Kingdom	Peltation
MON	Alismatales	Araceae	*Colocasia esculenta* (L.) Schott	Asia	SE Asia	h p	t	trop, boreostrop, m	Pal	e
MON	Alismatales	Araceae	*Colocasia fallax* Schott	Asia	SE Asia	h p	t	m, boreostrop	Pal	e
MON	Alismatales	Araceae	*Colocasia hassanii* H. Ara	Asia	Bangladesh	h p	t	boreostrop	Pal	e
MON	Alismatales	Araceae	*Colocasia mannii* Hook. f.	Asia	Bangladesh, India	h p	t	boreostrop	Pal	e
MON	Alismatales	Araceae	*Remusatia hookeriana* Schott	Asia	China, India, Myanmar, Nepal, Thailand	h p	t	m, boreostrop, trop	Hol, Pal	e
MON	Alismatales	Araceae	*Remusatia pumila* (D.Don) H.Li and A.Hay	Asia	China, Bangladesh, Nepal, Thailand	h p	t	boreostrop	Pal	e
MON	Alismatales	Araceae	*Remusatia vivipara* (Roxb.) Schott	Africa, Asia	Africa, Australia, SE Asia	h p	t	strop, trop	Pal, Aus	e
MON	Alismatales	Araceae	*Remusatia yunnanensis* (H. Li and A. Hay) A. Hay	Asia	China, Taiwan	h p	t	m, boreostrop	Hol, Pal	e
MON	Alismatales	Araceae	*Steudnera assamica* Hook. f.	Asia	India	h p	t	boreostrop	Pal	e
MON	Alismatales	Araceae	*Steudnera colocasiifolia* K.Koch	Asia	SE Asia, China	h p	t	m, boreostrop	Hol, Pal	e
MON	Alismatales	Araceae	*Steudnera discolor* W. Bull	Asia	India, Bangladesh, Myanmar, Thailand	h p	t	boreostrop	Pal	e
MON	Alismatales	Araceae	*Steudnera kerrii* Gagnep.	Asia	China, Laos, Thailand, Vietnam	h p	t	boreostrop	Pal	e
MON	Alismatales	Araceae	*Steudnera virosa* (Roxb.) Prain	Asia	Bangladesh, India	h p	t	boreostrop, trop	Pal	e
MON	Alismatales	Araceae	*Xanthosoma peltatum* G.S. Bunting	South America	Venezuela	h p	t	trop	Neo	e
EUD	Proteales	Nelumbonaceae	*Nelumbo lutea* Willd.	North America	S USA	h p	a	boreostrop, m	Hol	c
EUD	Proteales	Nelumbonaceae	*Nelumbo nucifera* Gaertn.	Asia	India	h p	a	boreostrop	Pal	c
EUD	Ranunculales	Berberidaceae	*Podophyllum delavayi* Franch.	Asia	SC China	h p	t	m	Hol	e
EUD	Ranunculales	Berberidaceae	*Podophyllum glaucescens* J.M.H.Shaw	Asia	SE China	h p	t	m	Hol	e
EUD	Ranunculales	Berberidaceae	*Podophyllum peltatum* L.	North America	N America	h p	t	m, sm, temp, b	Hol	c
EUD	Ranunculales	Berberidaceae	*Podophyllum pleianthum* Hance	Asia	SE China, Taiwan	h p	t	m	Hol	c
EUD	Ranunculales	Berberidaceae	*Podophyllum versipelle* Hance	Asia	China, Vietnam	h p	t	boreostrop	Pal	c
EUD	Ranunculales	Menispermaceae	*Cissampelos glaberrima* A. St.-Hil.	South America	Tropical South America	h,l p	t	trop, austrostrop	Neo	e

Table 1. *Cont.*

Clade	Order	Family	Species	Continent	Distribution	Habit	Habitat	Floral Zone	Floristic Kingdom	Peltation
EUD	Ranunculales	Menispermaceae	*Cissampelos grandifolia* Triana and Planch.	North America, South America	C America, Tropical S America	h,l p	t	boreostrop, trop, austrostrop	Neo	e
EUD	Ranunculales	Menispermaceae	*Cissampelos hispida* Forman	Asia	Thailand	l p	t	boreostrop, trop	Pal	e
EUD	Ranunculales	Menispermaceae	*Cissampelos owariensis* P. Beauv. ex DC.	Africa	Tropical Africa	h,l p	t	boreostrop, trop	Pal	e
EUD	Ranunculales	Menispermaceae	*Cissampelos sympodialis* Eichler	South America	Brazil	h,l p	t	trop, austrostrop	Neo	e
EUD	Ranunculales	Menispermaceae	*Coscinium blumeanum* Miers ex Hook.f. and Thomson	Asia	Malaysia, Thailand, Vietnam	l p	t	boreostrop, trop	Pal	e
EUD	Ranunculales	Menispermaceae	*Cyclea cauliflora* Merr.	Asia	Philippines, Sulawesi	h,l p	t	boreostrop, trop	Pal	e
EUD	Ranunculales	Menispermaceae	*Cyclea debiliflora* Miers	Asia	China, India, Vietnam	h,l p	t	m, boreostrop, trop	Pal	e
EUD	Ranunculales	Menispermaceae	*Cyclea fissicalyx* Dunn	Asia	India	h,l p	t	boreostrop, trop	Pal	e
EUD	Ranunculales	Menispermaceae	*Cyclea hypoglauca* (Schauer) Diels	Asia	China, Vietnam	h,l p	t	m, boreostrop, trop	Pal	e
EUD	Ranunculales	Menispermaceae	*Cyclea peltata* (Lam.) Hook. f. and Thomson	Asia	SE Asia, India	h,l p	t	boreostrop	Pal	e
EUD	Ranunculales	Menispermaceae	*Disciphania calocarpa* Standl.	North America	C America	l p	t	boreostrop	Neo	e
EUD	Ranunculales	Menispermaceae	*Disciphania contraversa* Barneby	South America	Brazil	- p	t	austrostrop	Neo	e
EUD	Ranunculales	Menispermaceae	*Disciphania hernandia* (Vell.) Barneby	South America	Brazil	h,l p	t	trop, austrostrop	Neo	e
EUD	Ranunculales	Menispermaceae	*Menispermum dauricum* DC.	Asia	China, Mongolia, Russia, Japan, Korea	h,l p	t	b, temp, sm, m, boreostrop	Hol, Pal	e
EUD	Ranunculales	Menispermaceae	*Perichasma laetificata* Miers	Africa	WC Africa	- -	t	trop	Pal	e
EUD	Ranunculales	Menispermaceae	*Stephania abyssinica* (Quart.-Dill. and A.Rich.) Walp.	Africa	Africa	h p	t	strop, trop	Pal	e
EUD	Ranunculales	Menispermaceae	*Stephania brevipes* Craib	Asia	Thailand, Vietnam	h p	t	boreostrop, trop	Pal	e
EUD	Ranunculales	Menispermaceae	*Stephania delavayi* Diels	Asia	China, Myanmar	h p	t	boreostrop	Pal	e

Table 1. *Cont.*

Clade	Order	Family	Species	Continent	Distribution	Habit	Habitat	Floral Zone	Floristic Kingdom	Peltation
EUD	Ranunculales	Menispermaceae	*Stephania japonica* (Thunb.) Miers	Asia, Australia and Oceania	Tropical Asia, Subtropical Asia, Australia	h p	t	boreostrop, trop, austrostrop	Pal, Aus	e
EUD	Ranunculales	Menispermaceae	*Stephania venosa* Spreng.	Asia	SE Asia	h p	t	boreostrop, trop	Pal	e
EUD	Ranunculales	Ranunculaceae	*Asteropyrum cavaleriei* H.Lév. and Vaniot	Asia	China	h p	t	m	Hol	e
EUD	Ranunculales	Ranunculaceae	*Asteropyrum peltatum* (Franch.) J.R.Drumm. and Hutch.	Asia	China, Myanmar	h p	t	boreostrop	Pal	e
EUD	Ranunculales	Ranunculaceae	*Peltocalathos baurii* (MacOwan) Tamura	Africa	South Africa, Lesotho, Eswatini	h p	t	austr	Cap	c
EUD	Ranunculales	Ranunculaceae	*Ranunculus clypeatus* (Ulbr.) Lourteig	South America	Peru	h p	t	trop	Neo	e
EUD	Ranunculales	Ranunculaceae	*Ranunculus lyallii* Hook.f.	Australia	New Zealand	h p	t	austr	Ant	c
EUD	Ranunculales	Ranunculaceae	*Thalictrum ichangense* Lecoy. ex Oliv.	Asia	China, Korea, Vietnam	h p	t	boreostrop, m	Hol, Pal	e
EUD	Ranunculales	Ranunculaceae	*Thalictrum pringlei* S. Watson	North America	Mexico	h p	t	boreostrop	Neo	e
EUD	Ranunculales	Ranunculaceae	*Thalictrum pseudoichangense* Q.E. Yang and G.H. Zhu	Asia	China	h p	t	m, boreostrop	Hol, Pal	e
EUD	Ranunculales	Ranunculaceae	*Thalictrum roseanum* B.Boivin	North America	Mexico	h p	t	boreostrop	Neo	e
EUD-C	Apiales	Apiaceae	*Klotzschia brasiliensis* Cham.	South America	Brazil	h −	t	austrostrop	Neo	e
EUD-C	Apiales	Apiaceae	*Klotzschia glaziovii* Urb.	South America	Brazil	l p	t	austrostrop	Neo	e
EUD-C	Apiales	Apiaceae	*Klotzschia rhizophylla* Urb.	South America	Brazil	h −	t	austrostrop	Neo	e
EUD-C	Apiales	Apiaceae	*Petagnaea gussonei* (Spreng.) Rauschert	Europe	Sicilia	h p	t	m	Hol	c
EUD-C	Apiales	Araliaceae	*Hydrocotyle bonariensis* Lam.	North America, South America	C America, S America	h p	t	boreostrop, trop	Neo	c
EUD-C	Apiales	Araliaceae	*Hydrocotyle pusilla* R.Br. ex Rich.	North America, South America	C America, S America	h p	t	boreostrop, trop	Neo	c
EUD-C	Apiales	Araliaceae	*Hydrocotyle umbellata* L.	North America, South America	C America, S America	h p	t	boreostrop, trop	Neo	c
EUD-C	Apiales	Araliaceae	*Hydrocotyle vulgaris* L.	Europe	Europe	h p	a	m, sm, temp, b	Hol	c
EUD-C	Apiales	Araliaceae	*Hydrocotyle yanghuangensis* (Hieron.) Mathias	South America	Ecuador	h p	t	trop	Neo	e

Table 1. *Cont.*

Clade	Order	Family	Species	Continent	Distribution	Habit	Habitat	Floral Zone	Floristic Kingdom	Peltation
EUD-C	Apiales	Araliaceae	*Oplopanax japonicus* Nakai	Asia	Japan	l p	t	t, sm	Hol	e
EUD-C	Apiales	Araliaceae	*Schefflera actinophylla* (Endl.) Harms	Australia and Oceania	Australia, New Guinea	l p	t	trop, austrostrop	Pal, Aus	e
EUD-C	Apiales	Araliaceae	*Schefflera arboricola* (Hayata) Merr.	Asia	Taiwan	l p	t	m	Hol	e
EUD-C	Apiales	Araliaceae	*Schefflera digitata* J.R. Forst. and G. Forst.	Australia and Oceania	New Zealand	l p	t	austrostrop, austr	Aus	e
EUD-C	Asterales	Asteraceae	*Ligularia nelumbifolia* Hand.-Mazz.	Asia	China	h p	t	sm, m, boreostrop	Hol, Pal	e
EUD-C	Asterales	Asteraceae	*Prenanthes subpeltata* Stebbins	Africa	Ethiopia, Kenya, Rwanda, Democratic Republic of Congo	h p	t	boreostrop, trop	Pal	e
EUD-C	Asterales	Asteraceae	*Psacalium laxiflorum* Benth.	North America	Mexico	h p	t	boreostrop	Neo	c
EUD-C	Asterales	Asteraceae	*Psacalium megaphyllum* Rydb.	North America	Mexico	h p	t	boreostrop	Neo	e
EUD-C	Asterales	Asteraceae	*Psacalium peltatum* Cass.	North America	Mexico	h p	t	boreostrop	Neo	c
EUD-C	Asterales	Asteraceae	*Psacalium pinetorum* (Standl. and Steyerm.) Cuatrec.	North America	Guatemala	h p	t	boreostrop	Neo	e
EUD-C	Asterales	Asteraceae	*Psacalium putlanum* B.L. Turner	North America	Mexico	h p	t	boreostrop	Neo	e
EUD-C	Asterales	Asteraceae	*Roldana chapalensis* (S. Watson) H. Rob. and Bretell	North America	Mexico	l p	t	boreostrop	Neo	e
EUD-C	Asterales	Asteraceae	*Roldana heterogama* (Benth.) H.Rob. and Brettell	North America	Costa Rica, Guatemala, Mexico, Panama	h,l p	t	boreostrop	Neo	e
EUD-C	Asterales	Asteraceae	*Roldana subpeltata* (Sch. Bip.) H. Rob and Brettell	North America	Mexico	h,l p	t	boreostrop	Neo	e
EUD-C	Asterales	Asteraceae	*Senecio oxyriifolius* DC.	Africa	C Africa, S Africa	h p	t	austr, austrostrop, trop	Cap, Pal	e
EUD-C	Asterales	Asteraceae	*Syneilesis aconitifolia* (Bunge) Maxim.	Asia	China, Japan, Korea, Russia	h p	t	t, sm, m, boreostrop	Hol, Pal	c
EUD-C	Asterales	Asteraceae	*Syneilesis palmata* (Thunb.) Maxim.	Asia	Korea, Japan	h p	t	sm	Hol	c
EUD-C	Brassicales	Caricaceae	*Jacaratia digitata* (Poepp. and Endl.) Solms	South America	Bolivia, Brazil, Ecuador, Venezuela	l p	t	strop, trop	Neo	e
EUD-C	Brassicales	Caricaceae	*Jacaratia spinosa* (Aubl.) A.DC.	South America	C America, S America	l p	t	strop, trop	Neo	e
EUD-C	Brassicales	Tropaeolaceae	*Tropaeolum ciliatum* Ruiz and Pav.	South America	Chile	h p	t	austrostrop	Neo, Ant	e
EUD-C	Brassicales	Tropaeolaceae	*Tropaeolum majus* L.	South America	Peru	h p	t	austrostrop	Neo	e
EUD-C	Brassicales	Tropaeolaceae	*Tropaeolum minus* L.	South America	Ecuador, Peru	h a	t	austrostrop, trop	Neo	e

Table 1. *Cont.*

Clade	Order	Family	Species	Continent	Distribution	Habit	Habitat	Floral Zone	Floristic Kingdom	Peltation
EUD-C	Brassicales	Tropaeolaceae	*Tropaeolum pentaphyllum* Lam.	South America	Argentina, Brazil, Paraguay, Uruguay	h p	t	austrostrop	Neo	e
EUD-C	Brassicales	Tropaeolaceae	*Tropaeolum tuberosum* Ruiz and Pav.	South America	Bolivia, Colombia, Ecuador, Peru	h p	t	austrostrop, trop	Neo	e
EUD-C	Caryophyllales	Polygonaceae	*Coccoloba acapulcensis* Standl.	North America	C America	l p	t	boreostrop	Neo	e
EUD-C	Caryophyllales	Polygonaceae	*Coccoloba tiliacea* Lindau	South America	Argentina, Bolivia	l p	t	trop, austrostrop, austr	Neo	e
EUD-C	Caryophyllales	Polygonaceae	*Persicaria perfoliata* (L.) Gross	Europe, Asia	Turkey, India, E Asia, SE Asia	h a	t	b, temp, sm, m, boreostrop, trop	Hol, Pal	e
EUD-C	Cornales	Loasaceae	*Nasa peltata* (Spruce ex Urb. and Gilg) Weigend	South America	Ecuador, Peru	h -	t	trop	Neo	e
EUD-C	Cornales	Loasaceae	*Nasa peltiphylla* (Weigend) Weigend	South America	Colombia, Ecuador	- -	t	trop	Neo	e
EUD-C	Cucurbitales	Begoniaceae	*Begonia caroliniifolia* Regel	North America	Mexico, Guatemala, Honduras	h p	t	boreostrop	Neo	e
EUD-C	Cucurbitales	Begoniaceae	*Begonia goegoensis* N.E.Br.	Asia	Sumatra	h p	t	trop	Pal	e
EUD-C	Cucurbitales	Begoniaceae	*Begonia kellermannii* C.DC.	South America	S America	h p	t	trop	Neo	e
EUD-C	Cucurbitales	Begoniaceae	*Begonia nelumbiifolia* Cham. and Schltdl.	South America	S America	h p	t	austrostrop	Neo	e
EUD-C	Cucurbitales	Begoniaceae	*Begonia sudjanae* C.-A. Jansson	Asia	Sumatra	h p	t	trop	Pal	e
EUD-C	Ericales	Balsaminaceae	*Impatiens begonioides* Eb. Fisch. and Raheliv.	Africa	Madagascar	h -	t	trop, austrostrop	Pal	e
EUD-C	Fabales	Fabaceae	*Lupinus angustifolius* L.	Europe, Africa, Asia	Mediterranean area, China	h a	t	sm, m	Hol, Pal	e
EUD-C	Fabales	Fabaceae	*Lupinus arboreus* Sims	North America	W USA	h,l p	t	sm, m	Hol	e
EUD-C	Fabales	Fabaceae	*Lupinus digitatus* Forssk.	Africa	N Africa, Senegal	h a	t	m, boreostrop	Hol, Pal	e
EUD-C	Fabales	Fabaceae	*Lupinus gussoneanus* J. Agardh	Europe	Mediterranean area	h a	t	sm, m	Hol	e
EUD-C	Fabales	Fabaceae	*Lupinus mutabilis* Sweet	South America	Bolivia, Colombia, Ecuador, Peru, Venezuela	h,l p	t	trop	Neo	e
EUD-C	Gentianales	Apocynaceae	*Hoya imbricata* Decne.	Asia	Philippines, Sulawesi	h p	t	boreostrop, trop	Pal	c

Table 1. *Cont.*

Clade	Order	Family	Species	Continent	Distribution	Habit	Habitat	Floral Zone	Floristic Kingdom	Peltation
EUD-C	Gentianales	Apocynaceae	*Macropharynx abnorma* J.F. Morales, M.E. Endress and Liede	South America	Peru	h,l p	t	trop	Neo	e
EUD-C	Gentianales	Apocynaceae	*Macropharynx conflictiva* (J.F. Morales) J.F. Morales, M.E. Endress and Liede	South America	Peru	h,l p	t	trop	Neo	e
EUD-C	Gentianales	Apocynaceae	*Macropharynx gigantea* (Woodson) J.F. Morales, M.E. Endress and Liede	South America	Peru, Bolivia	h,l p	t	trop, austrostrop	Neo	e
EUD-C	Gentianales	Apocynaceae	*Macropharynx macrocalyx* (Müll. Arg.) J.F. Morales, M.E. Endress and Liede	South America	Brazil, Paraguay	h,l p	t	trop, austrostrop	Neo	e
EUD-C	Gentianales	Apocynaceae	*Macropharynx peltata* (Vell.) J.F. Morales, M.E. Endress and Liede	South America	Argentina, Brazil, Paraguay	h,l p	t	trop, austrostrop	Neo	e
EUD-C	Gunnerales	Gunneraceae	*Gunnera antioquensis* L.E. Mora	South America	Colombia	h p	t	trop	Neo	e
EUD-C	Gunnerales	Gunneraceae	*Gunnera brephogea* Linden and André	South America	Colombia, Ecuador, Peru	h p	t	trop	Neo	c
EUD-C	Gunnerales	Gunneraceae	*Gunnera peltata* Phil.	South America	Juan Fernández Is.	h p	t	austr	Ant	c
EUD-C	Gunnerales	Gunneraceae	*Gunnera quitoensis* L.E. Mora	South America	Ecuador	h p	t	trop	Neo	e
EUD-C	Gunnerales	Gunneraceae	*Gunnera silvioana* L.E. Mora	South America	Colombia, Ecuador	h p	t	trop	Neo	e
EUD-C	Lamiales	Gesneriaceae	*Cyrtandra toviana* F. Br.	Australia and Oceania	Marquesas Islands	l p	t	trop	Pal	e
EUD-C	Lamiales	Gesneriaceae	*Cyrtandra wawrae* C.B. Clarke	Australia and Oceania	Hawaii	- -	t	trop	Pal	e
EUD-C	Lamiales	Gesneriaceae	*Drymonia peltata* (Oliv.) H.E. Moore	North America	Costa Rica	- -	t	boreostrop	Neo	e
EUD-C	Lamiales	Gesneriaceae	*Metapetrocosmea peltata* (Merr. and Chun) W.T. Wand	Asia	China	h p	t	boreostrop	Pal	e
EUD-C	Lamiales	Gesneriaceae	*Paraboea peltifolia* D. Fang and L. Zeng	Asia	China	h p	t	m, boreostrop	Hol, Pal	e
EUD-C	Lamiales	Gesneriaceae	*Paraboea yunfuensis* F. Wen and Y.G. Wie	Asia	China	h p	t	m, boreostrop	Hol, Pal	e
EUD-C	Lamiales	Gesneriaceae	*Petrocosmea huanjiangensis* Yan Liu and W.B. Xu	Asia	China	h p	t	m, boreostrop	Hol, Pal	e
EUD-C	Lamiales	Gesneriaceae	*Petrocosmea pubescens* D.J. Middleton and Triboun	Asia	Thailand	h p	t	boreostrop, trop	Pal	e

Clade	Order	Family	Species	Continent	Distribution	Habit	Habitat	Floral Zone	Floristic Kingdom	Peltation
EUD-C	Lamiales	Gesneriaceae	*Sinningia tuberosa* (Mart.) H.E. Moore	South America	Brazil	h	p	t	austrostrop	Neo	e
EUD-C	Lamiales	Gesneriaceae	*Streptocarpus mandrerensis* Humbert	Africa	Madagascar	h	–	t	trop, austrostrop	Pal	e
EUD-C	Lamiales	Gesneriaceae	*Streptocarpus peltatus* Randrian., Phillipson, Lowry and Mich. Möller	Africa	Madagascar	h	p	t	trop, austrostrop	Pal	e
EUD-C	Lamiales	Gesneriaceae	*Trichodrymonia peltatifolia* (J.L. Clark and M.M. More) M.M. Mora and J.L. Clark	North America	Panama	h	–	t	boreostrop	Neo	e
EUD-C	Lamiales	Lentibulariaceae	*Utricularia nelumbifolia* Gardn.	South America	Brazil	h	p	a	austrostrop	Neo	c
EUD-C	Lamiales	Lentibulariaceae	*Utricularia pubescens* Sm.	Africa, South America, Asia	S America, Africa, India	h	a	t	strop, trop	Neo, Pal	c
EUD-C	Malpighiales	Euphorbiaceae	*Endospermum moluccanum* (Teijsm. and Binn.) Kurz	Asia, Australia and Oceania	Indonesia, New Guinea, Solomon Islands	l	p	t	trop	Pal	e
EUD-C	Malpighiales	Euphorbiaceae	*Endospermum peltatum* Merr.	Asia	SE Asia	l	p	t	trop	Pal	e
EUD-C	Malpighiales	Euphorbiaceae	*Homalanthus grandifolius* Ridl.	Asia	Borneo, Sumatra	l	p	t	trop	Pal	e
EUD-C	Malpighiales	Euphorbiaceae	*Homalanthus macradenius* Pax and K. Hoffm.	Asia	Philippines	l	p	t	boreostrop, trop	Pal	e
EUD-C	Malpighiales	Euphorbiaceae	*Jatropha hernandiifolia* Vent.	North America	Dominican Republic, Haiti, Puerto Rico	l	p	t	boreostrop	Neo	e
EUD-C	Malpighiales	Euphorbiaceae	*Jatropha nudicaulis* Benth.	South America	Colombia, Ecuador	l	p	t	trop	Neo	e
EUD-C	Malpighiales	Euphorbiaceae	*Jatropha peltata* Sessé	North America	Mexico	h,l	p	t	boreostrop	Neo	e
EUD-C	Malpighiales	Euphorbiaceae	*Jatropha podagrica* Hook.	North America	C America	h,l	p	t	trop	Neo	e
EUD-C	Malpighiales	Euphorbiaceae	*Jatropha weberbaueri* Pax and K. Hoffm.	South America	Peru	–		t	trop	Neo	e
EUD-C	Malpighiales	Euphorbiaceae	*Macaranga bancana* (Miq.) Müll.Arg.	Asia	Thailand, Malesia	l	p	t	boreostrop, trop	Pal	e
EUD-C	Malpighiales	Euphorbiaceae	*Macaranga cuspidata* Boivin ex Baill.	Africa	Madagascar	l	p	t	trop, austrostrop	Pal	e
EUD-C	Malpighiales	Euphorbiaceae	*Macaranga magna* Turrill	Australia and Oceania	Fiji	l	p	t	austrostrop	Aus	e
EUD-C	Malpighiales	Euphorbiaceae	*Macaranga tanarius* (L.) Mull.Arg.	Asia, Australia and Oceania	SE Asia, Australia	l	p	t	strop, trop	Pal, Aus	e
EUD-C	Malpighiales	Euphorbiaceae	*Macaranga thompsonii* Merr.	Australia and Oceania	Mariana Islands	l	p	t	trop	Pal	e

Table 1. *Cont.*

Clade	Order	Family	Species	Continent	Distribution	Habit	Habitat	Floral Zone	Floristic Kingdom	Peltation
EUD-C	MalpighialesEuphorbiaceae	*Mallotus floribundus* (Blume) Müll. Arg.	Asia	SE Asia	l	p t	boreostrop, trop	Pal	e
EUD-C	MalpighialesEuphorbiaceae	*Mallotus lackeyi* Elmer	Asia	Borneo, Philippines	l	p t	trop	Pal	e
EUD-C	MalpighialesEuphorbiaceae	*Mallotus peltatus* (Geiseler) Müll. Arg.	Asia	China, India, SE Asia	l	p t	m, boreostrop, trop	Pal	e
EUD-C	MalpighialesEuphorbiaceae	*Mallotus surculosus* P.I. Forst.	Australia and Oceania	Australia	l	p t	austrostrop	Aus	e
EUD-C	MalpighialesEuphorbiaceae	*Mallotus thorelii* Gagnep.	Asia	Cambodia, China, Laos, Thailand, Vietnam	l	p t	m, boreostrop, trop	Pal	e
EUD-C	MalpighialesEuphorbiaceae	*Manihot fabianae* M. Mend.	South America	Bolivia	l	p t	trop, austrostrop	Neo	e
EUD-C	MalpighialesEuphorbiaceae	*Manihot mirabilis* Pax	South America	Paraguay	l	p t	trop, austrostrop	Neo	e
EUD-C	MalpighialesEuphorbiaceae	*Manihot peltata* Pohl	South America	Brazil	l	p t	trop, austrostrop	Neo	e
EUD-C	MalpighialesEuphorbiaceae	*Megistostigma peltatum* (J.J. Sm.) Croizat	Asia	Java, Sumatra	l	p t	trop	Pal	e
EUD-C	MalpighialesEuphorbiaceae	*Meineckia peltata* (Hutch.) G.L. Webster	Africa	Madagascar	l	p t	trop, austrostrop	Pal	e
EUD-C	MalpighialesEuphorbiaceae	*Ricinus communis* L.	Africa	Eritrea, Ethiopia, Somalia	h,l	a,p t	boreostrop	Pal	e
EUD-C	Malpighiales Passifloraceae	*Adenia penangiana* (Wall. ex G. Don) W.J. de Wilde	Asia	SE Asia	h	p t	boreostrop, trop	Pal	e
EUD-C	Malpighiales Passifloraceae	*Passiflora coriacea* Juss.	South America	C America, S America	h	p t	boreostrop, trop	Neo	e
EUD-C	Malpighiales Passifloraceae	*Passiflora guatemalensis* S. Watson	North America, South America	C America, Colombia, Venezuela	h	p t	boreostrop, trop	Neo	e
EUD-C	Malpighiales Passifloraceae	*Passiflora rubrotincta* Killip	South America	Bolivia	h	p t	trop, austrostrop	Neo	e
EUD-C	Malpighiales Passifloraceae	*Passiflora spectabilis* Killip	South America	Peru	h	p t	trop	Neo	e
EUD-C	MalpighialesPhyllanthaceae	*Astrocasia peltata* Standl.	North America	Costa Rica, Mexico	l	p t	boreostrop	Neo	e
EUD-C	Malpighiales Salicaceae	*Xylosma peltata* (Sleumer) Lescot	Australia and Oceania	New Caledonia	l	p t	austrostrop	Aus	e
EUD-C	Malvales Dipterocarpaceae	*Shorea peltata* Symington	Asia	Borneo, Malaysia, Sumatra	l	p t	trop	Pal	e

Table 1. *Cont.*

Clade	Order	Family	Species	Continent	Distribution	Habit	Habitat	Floral Zone	Floristic Kingdom	Peltation
EUD-C	Malvales	Malvaceae	*Brownlowia ferruginea* Kosterm.	Asia	Borneo	l p	t	trop	Pal	e
EUD-C	Malvales	Malvaceae	*Brownlowia havilandii* Stapf	Asia	Borneo	l p	t	trop	Pal	e
EUD-C	Malvales	Malvaceae	*Brownlowia helferiana* Pierre	Asia	Malaysia, Myanmar, Thailand	l p	t	boreostrop, trop	Pal	e
EUD-C	Malvales	Malvaceae	*Brownlowia peltata* Benth.	Asia	Borneo, Myanmar, Thailand, Vietnam	l p	t	boreostrop, trop	Pal	e
EUD-C	Malvales	Malvaceae	*Brownlowia stipulata* Kosterm.	Asia	Borneo	l p	t	trop	Pal	e
EUD-C	Malvales	Malvaceae	*Pterospermum acerifolium (L.) Willd.*	Asia	India, SE Asia	h p	t	boreostrop, trop	Pal	e
EUD-C	Myrtales	Melastomataceae	*Catanthera peltata* M.P. Nayar	Asia	Borneo	-	t	trop	Pal	e
EUD-C	Myrtales	Melastomataceae	*Conostegia peltata* (Almeda) Kriebel	North America	Panama	l p	t	trop	Neo	e
EUD-C	Myrtales	Melastomataceae	*Gravesia peltata* H. Perrier	Africa	Madagascar	h p	t	trop, austrostrop	Pal	e
EUD-C	Myrtales	Melastomataceae	*Leandra peltata* Wurdack	South America	Peru	l p	t	trop	Neo	e
EUD-C	Myrtales	Melastomataceae	*Phyllagathis beccariana* (Cogn.) M.P. Nayar	Asia	Borneo	h p	t	trop	Pal	e
EUD-C	Myrtales	Melastomataceae	*Phyllagathis peltata* Stapf ex Ridl.	Asia	Borneo	h p	t	trop	Pal	e
EUD-C	Oxalidales	Oxalidaceae	*Oxalis articulata* Savigny	South America	Argentina, Brazil, Uruguay	h p	t	austrostrop	Neo	c
EUD-C	Oxalidales	Oxalidaceae	*Oxalis bowiei* W.T. Aiton ex G.Don	Africa	South Africa	h p	t	austr	Cap	c
EUD-C	Oxalidales	Oxalidaceae	*Oxalis decaphylla* Kunth	North America	Mexico	h p	t	boreostrop	Neo	c
EUD-C	Oxalidales	Oxalidaceae	*Oxalis leucolepis* Diels	Asia	China, Nepal	h p	t	boreostrop, m	Pal	c
EUD-C	Oxalidales	Oxalidaceae	*Oxalis triangularis* A.St.-Hil.	South America	South America	h p	t	austrostrop, trop	Neo	c
EUD-C	Rosales	Moraceae	*Dorstenia belizensis* C.C. Berg	North America	Belize	h p	t	boreostrop	Neo	e
EUD-C	Rosales	Moraceae	*Dorstenia erythrandra* C. Wright ex Griseb.	North America	Cuba, Dominican Republic, Haiti	h p	t	boreostrop	Neo	e
EUD-C	Rosales	Moraceae	*Dorstenia jamaicensis* Britton	North America	Jamaica	h p	t	boreostrop	Neo	e
EUD-C	Rosales	Moraceae	*Dorstenia nummularia* Urb. and Ekman	North America	Cuba	h p	t	boreostrop	Neo	e
EUD-C	Rosales	Moraceae	*Dorstenia peltata* Spreng.	North America	Cuba, Dominican Republic	h p	t	boreostrop	Neo	e
EUD-C	Rosales	Urticaceae	*Cecropia albicans* Trécil	South America	Peru	l p	t	trop	Neo	e
EUD-C	Rosales	Urticaceae	*Cecropia distachya* Huber	South America	N South America	l p	t	trop, austrostrop	Neo	e

Table 1. *Cont.*

Clade	Order	Family	Species	Continent	Distribution	Habit	Habitat	Floral Zone	Floristic Kingdom	Peltation
E-JD-C	Rosales	Urticaceae	*Cecropia elongata* Rusby	South America	Bolivia	l	p	t	trop, austrostrop	Neo	e
E-JD-C	Rosales	Urticaceae	*Cecropia latiloba* Miq.	South America	N South America	l	p	t	trop, austrostrop	Neo	e
E-JD-C	Rosales	Urticaceae	*Cecropia peltata* L.	South America	C America	l	p	t	boreostrop, trop	Neo	e
E-JD-C	Rosales	Urticaceae	*Dendrocnide moroides* (Wedd.) Chew	Asia, Australia and Oceania	Australia, Lesser Sunda Islands, Vanuatu	l	p	t	trop, austrostrop, austr	Pal, Aus	e
E-JD-C	Rosales	Urticaceae	*Dendrocnide peltata* (Blume) Miq.	Asia, Australia and Oceania	New Guinea, Malesia	l	p	t	trop	Pal	e
E-JD-C	Rosales	Urticaceae	*Elatostema muluense* Rodda and A.K. Monro	Asia	Borneo	h	p	t	trop	Pal	e
E-JD-C	Rosales	Urticaceae	*Elatostema peltifolium* (Ridl.) H.J.P. Winkl.	Australia and Oceania	New Guinea	h	-	t	trop	Pal	e
E-JD-C	Rosales	Urticaceae	*Musanga cecropioides* R. Br. ex Tedlie	Africa	W and C Africa	l	p	t	trop	Pal	c
E-JD-C	Rosales	Urticaceae	*Musanga leo-errerae* Hauman and J. Léonard	Africa	C Africa	l	p	t	trop	Pal	c
E-JD-C	Rosales	Urticaceae	*Pilea nonggangensis* Y.G. Wei, L.F. Fu and A.K. Monro	Asia	China	h	p	t	m, boreostrop	Hol, Pal	e
E-JD-C	Rosales	Urticaceae	*Pilea panzhihuaensis* C.J. Chen, A.K. Monro and L. Chen	Asia	China	h	p	t	m, boreostrop	Hol, Pal	e
E-JD-C	Rosales	Urticaceae	*Pilea peltata* Hance	Asia	China, Vietnam	h	p	t	m, boreostrop, trop	Hol, Pal	e
E-JD-C	Rosales	Urticaceae	*Pilea peperomioides* Diels	Asia	China	h	p	t	m	Hol	e
E-JD-C	Rosales	Rosaceae	*Rubus peltatus* Maxim.	Asia	China, Japan	h	p	t	sm	Hol	e
E-JD-C	Saxifragales	Crassulaceae	*Kalanchoe beharensis* Drake	Africa	Madagascar	h	p	t	trop, austrostrop	Pal	e
E-JD-C	Saxifragales	Crassulaceae	*Kalanchoe nyikae* Engl.	Africa	Kenya, Tanzania	h	p	t	trop	Pal	e
E-JD-C	Saxifragales	Crassulaceae	*Umbilicus botryoides* Hochst. ex A.Rich.	Africa	Africa	h	p	t	boreostrop, trop	Pal	c
E-JD-C	Saxifragales	Crassulaceae	*Umbilicus horizontalis* DC.	Africa, Europe	S Europe, N Africa	h	p	t	m	Hol	c
E-JD-C	Saxifragales	Crassulaceae	*Umbilicus luteus* Webb and Berthel.	Europe	Mediterranean region	h	p	t	m, sm	Hol	c

Table 1. *Cont.*

Clade	Order	Family	Species	Continent	Distribution	Habit	Habitat	Floral Zone	Floristic Kingdom	Peltation
EUD-C	Saxifragales	Crassulaceae	*Umbilicus rupestris* (Salisb.) Dandy	Africa, Europe	Europe, Africa	h	p	t	m, sm	Hol	c
EUD-C	Saxifragales	Saxifragaceae	*Astilboides tabularis* Engl.	Asia	China	h	p	t	sm, temp	Hol	c
EUD-C	Saxifragales	Saxifragaceae	*Chrysosplenium peltatum* Turcz.	Asia	Mongolia, Russia	h	p	t	t, sm, m	Hol	e
EUD-C	Saxifragales	Saxifragaceae	*Darmera peltata* (Torr.) Voss	North America	USA	h	p	t	m	Hol	c
EUD-C	Saxifragales	Saxifragaceae	*Peltoboykinia tellimoides* (Maxim.) H. Hara	Asia	China, Japan	h	p	t	t, sm, m	Hol, Pal	e
EUD-C	Saxifragales	Saxifragaceae	*Peltoboykinia watanabei* (Yatabe) H. Hara	Asia	Japan	h	p	t	t, sm, m	Hol	e
EUD-C	Saxifragales	Saxifragaceae	*Rodgersia aesculifolia* Batalin	Asia	China, Mongolia	h	p	t	sm, m, boreostrop	Hol, Pal	c
EUD-C	Saxifragales	Saxifragaceae	*Rodgersia podophylla* A.Gray	Asia	Japan, Korea	h	p	t	sm	Hol	c
EUD-C	Solanales	Convolvulaceae	*Decalobanthus elmeri* (Merr.) A.R. Simoes and Staples	Asia	Borneo	l	p	t	trop	Neo	e
EUD-C	Solanales	Convolvulaceae	*Decalobanthus peltatus* (L.) A.R. Simoes and Staples	Asia, Australia and Oceania	SE Asia, Australia, Madagascar	l	p	t	trop, austrostrop	Pal, Aus	e
EUD-C	Solanales	Solanaceae	*Nothocestrum peltatum* Skottsb.	Australia and Oceania	Hawaii	l	p	t	trop	Pal	e

The species are sorted by clade and order/family (alphabetically). A maximum of five species per genus are shown (for full list: see Supplementary Materials). Further information shown: distribution, habit, habitat, floristic zone and kingdom, peltation. Abbreviations: in clade: F (ferns), ANA (ANA grade), MAG (magnoliids), MON (monocots), EUD (eudicots), EUD-C (core eudicots), in distribution: N (northern), E (eastern), S (southern), W (western), C (central), in habit: h (herbaceous), l (lignified), a (annual), p (perennial), in habitat: a (aquatic), t (terrestrial), in floristic kingdom: Hol (Holarctic), Pal (Palaeotropics), Neo (Neotropics), Aus (Australis), Cap (Capensis), Ant (Antarctic), circpol (all continents in the floral zone), in floral zone: arct (arctic), b (boreal), temp (temperate), sm (submeridional), m (meridional), boreostrop (boreo-subtropical), strop (subtropical), austrostrop (austro-subtropical), trop (tropical), austr (austral), antarct (antarctic), in peltation: e (eccentric), c (central).

3.2. Leaf Anatomy, Morphology and Characteristics of Selected Peltate Plant Species

The leaf anatomy of 41 peltate leaved plant species from 18 different families was studied and categorised (Figure 2). For reasons of clarity, per family a maximum of two species were selected. Profiles of the excluded taxa can be found in the Supplementary Materials.

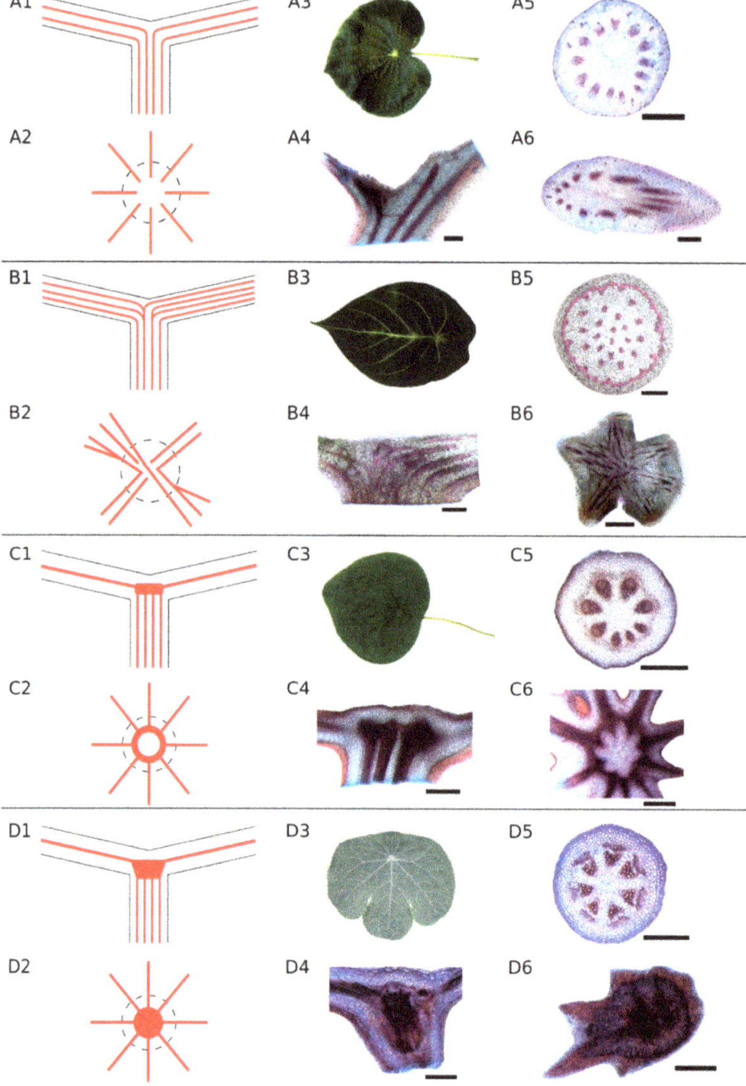

Figure 2. *Cont.*

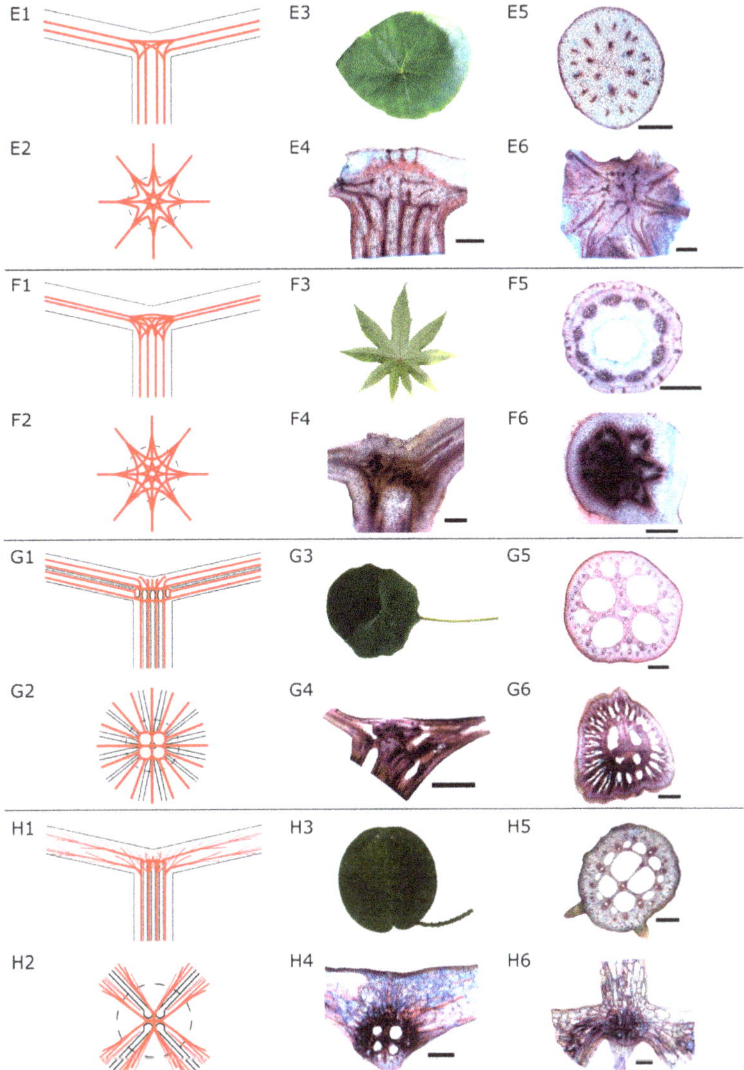

Figure 2. Types of strengthening structures in the petiole–lamina junction of peltate leaves (part 1). (**A**)—unbranched type (*Piper peltatum*, scale: 1 mm). (**B**)—monocot branching type (*Anthurium forgetii*, scale: 1 mm). (**C**)—ring-like structure (*Stephania delavayi*, scale: 500 µm). (**D**)—knot-like structure (*Tropaeolum tuberosum*, scale: (**D4,5**)—500 µm, (**D6**)—1 mm). (**E**)—net-like structure, intensity 1 (*Begonia nelumbiifolia*, scale: (**E4**), (**E6**)—2 mm, (**E5**)—1 mm). (**F**)—net-like structure, intensity 2 (*Ricinus communis*, scale: 1 mm). (**G**)—water plant radial branching type, (*Nelumbo nucifera*, scale: (**G4**)—5 mm, (**G5**)—1 mm, (**G6**)—2 mm). (**H**)—water plant diffuse branching type (*Victoria cruziana*, scale: (**H4**), (**H6**)—2 mm, (**H5**)—1 mm). 1—simplified illustration of strengthening structure type, longitudinal section of petiole and transition zone. 2—simplified illustration of strengthening structure type, top view of petiole (dashed line) and transition zone. 3—leaf morphology of peltate representative. 4—longitudinal section of the transition zone. 5—cross section of the petiole. 6—cross section of the transition zone.

3.2.1. ANA Grade

Nymphaeaceae Salisbury

Euryale ferox Salisb.—This species, also named Foxnut, is often cultivated for its seeds and fruits. Habit: rhizomatous aquatic plant with annual leaves; large round leaf plates (1.3–2.7 m) on the surface, prickly on the top and bottom; under water non-spiny cordiform leaves of 4–10 cm; leaves abaxial dark violet, adaxial green, peltate, central petiole insertion; petiole covered with spines; flower up to 5 cm in diameter; petals elongated lanceolate, purple–violet exterior and white interior. Distribution: native in India, China, Japan and some countries in Southeast Asia. Ecology: lakes and ponds [30,38,39]. Petiole anatomy: large aerenchyma tubes, scattered vascular bundles (number of vascular bundles: 42), layers of collenchyma tissue below epidermis. Anatomy of the transition zone: parallel fibre strands from the petiole branch out in the transition zone forming a diffuse net of fibres in different sizes, fibres running into the lamina follow main veins and become increasingly finer, air channels are visible. Additional data: see Table 2.

Victoria cruziana Orb.—It is cultivated as an ornamental plant. Habit: rhizomatous, ground rooted aquatic herb with annual branches and leaves; all parts prickly except upper side of the leaves and most surfaces of the flowers; leaves bright greenish, up to 2 m wide, vertically oriented leaf margins (20 cm high); lily pads emerge from a rhizome by a long, flexible petiole; flowers large, floating, first white, after pollination turn to light pink, anchored by a long stalk arising from the rhizome. Ecology: grows in slowly moving and shallow water. Distribution: Northeast Argentina, Paraguay and Bolivia [30]. Petiole anatomy: large aerenchyma tubes, scattered vascular bundles (number of vascular bundles: 12), layers of collenchyma tissue below epidermis (Figure 2H5). Anatomy of the transition zone: parallel fibre strands from the petiole branch out in the transition zone forming a diffuse net of fibres in different sizes, fibres running into the lamina follow main veins and become increasingly finer, air channels are visible (Figure 2H4,H6). Additional data: see Table 2.

Additional studied species: *Cabomba aquatica* Aubl., *Nymphaea colorata* Peter, *Nymphaea lotus* L., *Victoria amazonica* (Poepp.) J.C. Sowerby.

Table 2. Leaf characteristics of peltate plant species. The number of petiole vascular bundles (v.b.), ratio of strengthening tissue (only lignified tissue) in % in petiole cross section, strengthening structure, the ratio of petiole area to lamina area in % and water content in % of petiole, lamina in total, intercostal areas and venation of 41 peltate species were analysed

Species	Number of v.b.	Proportion of Strength. Tissue (in Petiole, %)	Strengthening Structure	Sample Size (n)	Petiole Area/Lamina Area (%)	Water Content Petiole (%)	Water Content Lamina (%)	Water Content Intercostal (%)	Water Content Venation (%)
Alocasia longiloba	53	2.7	Sb						
Amorphophallus konjac	180	10.3	Ub						
Anthurium forgetii	60	9.7	Sb						
Begonia kellermanii	11	5.4	Ns1	5	0.14	99.3	98.7	98.9	95.9
Begonia nelumbiifolia	34	8.0	Ns1	4	0.06	95.2	89.7	89.7	89.4
Begonia peltata	24	4.2	Ns1						
Begonia sudjanae				4	0.08	94.8	92.3	92.4	92.2
Cabomba aquatica	2	1 1	?	5	0.81	93.4	85.4	85.1	88.3
Caladium hybrid	50	3.3	Sb	3	0.04	94.5	87.6	86.4	91.8
Cecropia peltata	23	11.1	Ns2	8	0.03	89.6	83.6	78.7	90.6
Colocasia esculenta	90	4.1	Sb	2	0.05	95.6	89.8	88.8	92.2
Darmera peltata	24	8.8	Ub	2	0.12	88.9	83.7	82.6	87.1
Euryale ferox	42	6.4	Wdb	5	0.28	96.8	93.8	92.0	95.3
Hernandia nymphaeifolia	11	8.5	Sb	6	0.07	88.5	81.1	80.2	88.5
Hydrocotyle vulgaris	3	2.2	Ks	4	0.30	89.7	84.9	84.7	85.3
Jatropha podagrica	10	10.8	Ns2	2	0.08				
Kalanchoe nyikae	3	3.1	Sb						
Macaranga tanarius	27	18.8	Ns2	3	0.04	77.5	76.9	76.8	77.0
Marsilea strigosa	1	3.9	Sb						
Nelumbo nucifera	63	19 3	Wrb	5	0.09	91.0	88.0	87.5	89.2
Nymphaea colorata	21	5.7	Wrb						
Nymphaea lotus				1	0.10	97.3	91.5	90.9	92.4
Oxalis bowiei	7	3.0	Ub						
Passiflora coriacea	6	13.2	Ub	2	0.08	85.6	86.2	86.1	86.6
Peperomia argyreia	8	1.5	Sb						
Peperomia cyclaminoides	6	1.1	Sb	2	0.79	96.1	94.1	94.1	94.1
Peperomia sodiroi	7	4.3	Sb	3	0.25	93.5	92.9	92.6	89.8
Perichasma laetificata	8	18.8	Rs						
Pilea peperomioides	6	10.7	Ub	7	0.16	95.0	95.1	95.1	95.5

Table 2. *Cont.*

Species	Number of v.b.	Proportion of Strength. Tissue (in Petiole, %)	Strengthening Structure	Sample Size (n)	Petiole Area/Lamina Area (%)	Water Content Petiole (%)	Water Content Lamina (%)	Water Content Intercostal (%)	Water Content Venation (%)
Piper peltatum	13	8.2	Ub	3	0.03	91.7	85.0	83.8	87.8
Podophyllum hybrid	15	12.5	Ns1						
Remusatia vivipara	28	3.7	Sb	5	0.08	95.4	87.7	86.7	90.6
Ricinus communis	9	13.4	Ns2	4	0.05	86.4	77.6	76.7	81.0
Rodgersia podophylla	63	7.6	Ub	2	0.04	89.1	85.9	80.6	88.4
Schefflera arboricola	39	28.5	Ns1	1	0.05	90.6	75.8	75.6	76.7
Stephania delavayi	8	12.1	Rs	6	0.05	91.3	79.9	78.7	84.6
Syneilesis palmata	30	9.0	Sb	1	0.08				
Tropaeolum tuberosum	7	12.8	Ks	3	0.08	92.2	85.8	85.4	87.7
Umbilicus rupestris	8	6.8	Ub	1	0.44	97.5	97.0	97.0	96.9
Victoria amazonica	90	11.1	Wdb						
Victoria cruziana	12	7.9	Wdb	1	0.08				

Abbreviations for strengthening structures: Ub—unbranched, Sb—simple branching, Ns1—net-like structure, intensity 1, Ns2—net-like structure, intensity 2, Ks—knot-like structure, Rs—ring-like structure, Wdb—water plant diffuse branching type, Wrb—water plant radial branching type). Data shown in Figure 3A–D.

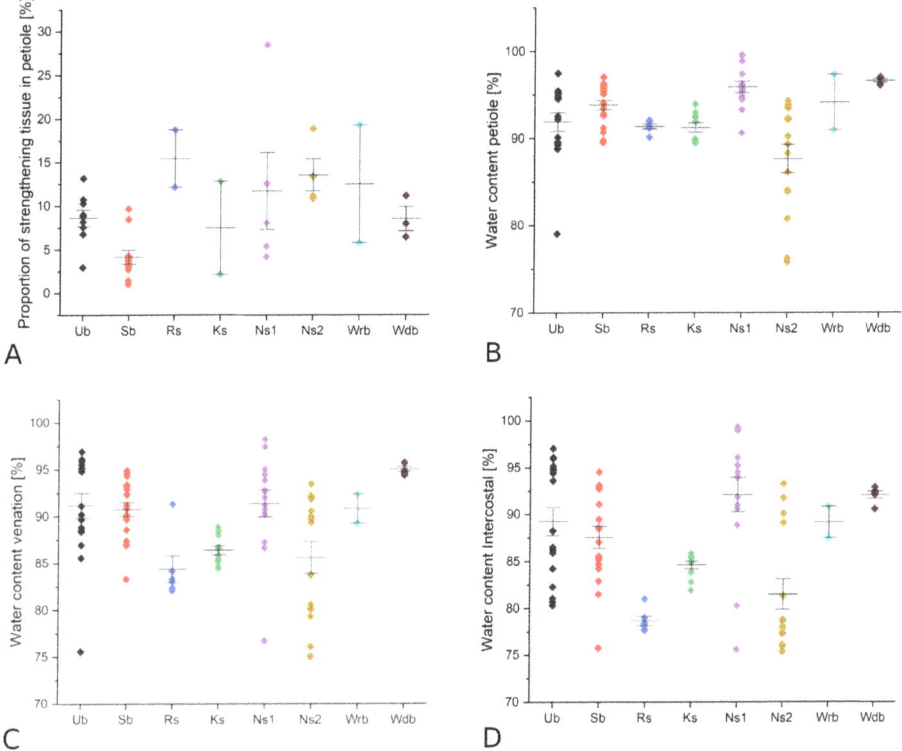

Figure 3. Proportion of (lignified) strengthening tissue in petiole (**A**) and water content of petiole (**B**), venation (**C**) and intercostal areas (**D**) in % of the different types of strengthening structures in peltate leaves (with mean and standard error). Abbreviations for strengthening structures: Ub—unbranched, Sb—simple branching, Ns1—net-like structure, intensity 1, Ns2—net-like structure, intensity 2, Ks—knot-like structure, Rs—ring-like structure, Wdb—water plant diffuse branching type, Wrb—water plant radial branching type.

3.2.2. Magnoliids

Hernandiaceae Blume

Hernandia nymphaeifolia (C. Presl) Kubitzki—Habit: tree or shrub, height: 3–20 m, spreading crown; petiole width: 1–3 cm, petiole length: 5–17 cm; leaves: 6–22 cm long, glabrous, peltate, ovate, entire margins, leaf base broadly rounded to rarely subcordate, leaf tip hastate, upper surface shiny green with eight to nine nerves; inflorescences with white- or cream-coloured flowers. Distribution: Southeast Asia, northern Australia and Madagascar. Ecology: sparse forests or bushland near sea level [30,40]. Petiole anatomy: circular arranged vascular bundles with sclerenchymatic caps on peripheral side (number of vascular bundles: 11), several layers of collenchyma under the epidermis. Anatomy of the transition zone: branching of some fibre strands, some fibre strands run unbranched into the apex of the leaf, fibre strands form a net-like structure in the transition zone and merge into large bundles following the leaf veins. Additional data: see Table 2.

Piperaceae Giseke

Peperomia sodiroi C. DC.—Habit: perennial, herbaceous, erect plant; stem glabrous; leaves oval, peltate, round at the base, tapered at the apical end, 5 cm wide, up to

6.5 cm long; petiole up to 9 cm long. Distribution: subtropical regions in Ecuador [30,41]. Petiole anatomy: circular arranged vascular bundles (number of vascular bundles: seven). Anatomy of the transition zone: fibre strands from the petiole spread out into lamina branching out minimally, forming a loose net-like structure and merging into large bundles following the leaf veins. Additional data: see Table 2.

Piper peltatum L.—Habit: shrub or subshrub, herbaceous, 0.5–2 m high; stem sparsely pilose or glabrous; leaves roundish-cordate, peltate, 15–30 cm wide, 20–40 cm long, apex pointed, base rounded to subcordate, glabrous, 10–15 veins radiating from the base of the petiole; petiole insertion point about a quarter to a third of the leaf length from the base, petioles 10–26 cm long, glabrous. Distribution: Mexico to Tropical America. Ecology: humid places at edges of meadows, forests, roadsides, stream edges, 0–500 (–800) m [42,43]. Petiole anatomy: circular arranged vascular bundles (number of vascular bundles: 13) around mucilage canal, collenchyma tissue below epidermis (Figure 2A5). Anatomy of the transition zone: fibre strands from the petiole spread unbranched into lamina (Figure 2A4,A6). Additional data: see Table 2.

Further studied species: *Peperomia argyreia* (Miq.) E. Morren., *Peperomia cyclaminoides* A.W. Hill.

3.2.3. Monocotyledons

Araceae Jussieu

Amorphophallus konjac K. Koch—Habit: slightly glossy, depressed globose tuber, approx. 30 cm in diameter; in vegetative phase producing a single leaf, base colour of the petiole dirty pink, covered with dark green and smaller white dots (100 cm × 8 cm), lamina forms leaflets, strongly divided up to 2 m in diameter, leaflets (3–10 cm × 2–6 cm) adaxially dull green, elliptically tapering; inflorescence mostly long-stemmed, up to 110 cm × 5 cm in size, at the base dirty pale-brownish with black–green spots or dirty pale-white grey with some scattered black–green spots, mottled purple–red at the edge, spadix during female anthesis produces strong smell of rotting meat and small, clear, slightly viscous droplets, flower time April. Distribution: China (Yunnan), introduced to Korea, Philippines, Thailand, Tibet, Vietnam, other Chinese regions. Ecology: glades or forest margins and thickets, in secondary forests, altitude 200–3000 m [30,44]. Petiole anatomy: scattered vascular bundles (number of vascular bundles: 180), collenchyma strands near the epidermis. Anatomy of the transition zone: fibre strands from the petiole spread unbranched into petiolules of the leaflets. Additional data: see Table 2.

Anthurium forgetii N.E.Br.—Habit: small erect herb, up to 38 cm high; petioles 15–25 cm long, erect, glabrous, leaves sagging, 25–35 cm long, 15–22 cm wide, peltate to elliptic–ovate, dark green, leaf margin entire, apex pointed, rounded at the base, upper side with a velvety sheen and pale green prominent nerves, underside pale green, sometimes with a slight violet tinge. Distribution: Colombia. Ecology: Andes, altitude 1100 m [30]. Petiole anatomy: central scattered vascular bundles, peripheral ring of vascular bundles (number of vascular bundles: 60), both with sclerenchyma bundle sheaths, sclerenchyma associated with peripheral vascular bundles forms closed ring (Figure 2B5). Anatomy of the transition zone: fibre strands from the petiole branch and proceed into leaf veins, some remain unbranched and follow the main vein (Figure 2B4,B6). Additional data: see Table 2.

Further studied species: *Alocasia longiloba* Miq., *Caladium* hybrid, *Remusatia vivipara* (Roxb.) Schott, *Colocasia esculenta* (L.) Schott.

3.2.4. Eudicotyledons

Araliaceae Jussieu

Hydrocotyle vulgaris L.—Habit: stoloniferous perennial plant; leaf lamina 1.5–4 cm in diameter, peltate, slightly notched, without incision at the base, petioles 3–18 cm long, about 1 mm thick, central leaf insertion. Distribution: generally oceanically distributed, Azores, Europe to Mediterranean. Ecology: fens, fen-meadows and swamps, dunes and damp hollows, occasionally in deeper water, also in wet heaths and ditches, generally calcifugal [30,45–47]. Petiole anatomy: three central vascular bundles surrounded by parenchyma. Anatomy of the transition zone: fibre strands from the petiole approach dense structure in transition zone, then separate into eight strands running into the leaf veins. Additional data: see Table 2.

Schefflera arboricola (Hayata) Merr.—Cultivated as an ornamental plant and used medicinally. Habit: shrub or rarely climbing up to 4 m height; petioles 6–20 cm long, pinnate leaves, leaves divide into 5–9 obovate–oblong to oblong or elliptical leaflets with 6–10 cm length and 1.5–3.5 cm width, leaflets are subleathery, glabrous on both sides, margins entire; inflorescence consists of terminal, glabrous corymbs with 3–8 cm. Distribution: Hainan (China), Taiwan, introduced in Florida (USA). Ecology: on banks of streams, in humid forests, altitude up to 900 m [30,48]. Petiole anatomy: vascular bundles arranged in two rings (number of vascular bundles: 39), vascular bundles in the inner ring arranged irregularly, sclerenchymatic bundle sheaths with large fibre caps envelop outer ring, layer of collenchyma under epidermis. Anatomy of the transition zone: transition zone between petiole and petiolules seems to be the predetermined breaking point, fibre strands branch out and then merge into petiolules. Additional data: see Table 2.

Asteraceae Bercht. and J. Presl

Syneilesis palmata (Thunb.) Maxim.—Habit: herbaceous perennial, robust, 60–75 cm height; leaf lamina up to 50 cm in diameter, peltate, palmately parted (7–9 lobes), ± orbicular, long petiole; flowers white, flowering time: summer. Distribution: Japan, Korea [30,49,50]. Petiole anatomy: vascular bundles arranged in 1–2 irregular rings, somewhat scattered, larger bundles peripheral, two central vascular bundles (number of vascular bundles: 30), large sclerenchymatic caps, several layers of collenchymatic tissue under epidermis. Anatomy of the transition zone: fibre strands from the petiole run mostly unbranched, with a few bundles splitting, through the transition zone into the lamina. Additional data: see Table 2.

Begoniaceae C. Agardh

Begonia peltata Otto and A. Dietr.—Habit: shrub with an erect stem; petioles 8–15 cm long, leaves approx. 15 cm long, taper from a broad base to a narrow tip, asymmetric, ovoid, peltate with an eccentric petiole insertion, palmately veined with just a few branches, petiole and leaves green with velutinous pubescence. Distribution: South Mexico to Honduras. Ecology: drought adapted plant, grows in dry canyons, in Guatemala it grows on the ground/on rocks in wet forests and thickets [30,51–53]. Petiole anatomy: vascular bundles arranged in two concentric rings (number of vascular bundles: 24), few layers of collenchymatic tissue below epidermis (Figure 2E5). Anatomy of the transition zone: fibre strands from petiole branch out in the transition zone forming a net and merge into fibre bundles, proceeding into the leaf veins (Figure 2E4,E6). Additional data: see Table 2.

Begonia nelumbiifolia Schltdl. and Cham—Habit: herbaceous perennial plant; leaf lamina glabrous, broadly ovate to peltate, base rounded, apex pointed, 15–40 cm long, 11–32 cm wide, seven to nine main nerves, petioles 11–63 cm long, 2–7 mm wide, hirsute; inflorescence rich in flowers, 7–28 cm in diameter, flowering time December to May. Distribution: Mexico to Colombia, introduced in Cuba, Dominican Republic and Puerto Rico. Ecology: altitude 100–400 m, along roadside verge, in Guatemala it grows in humid thickets and forests up to 1650 m [18,30,54–56]. Petiole anatomy: vascular bundles arranged in two concentric rings (number of vascular bundles: 34), few layers of collenchymatic tissue below epidermis. Anatomy of the transition zone: fibre strands from petiole branch out in the transition zone forming a net and merge into fibre bundles extending into the leaf veins. Additional data: see Table 2.

Further studied species: *Begonia kellermanii* C. DC., *Begonia sudjanae* C.-A. Jansson

Berberidaceae Jussieu

Podophyllum L.—Habit: deciduous, herbaceous perennials, 20–60 cm high, one leaf or flowering shoot per year emerge from a rhizome; leaves 10–38 cm, very diverse, but with a simple construction, petiole more or less centrally attached to the variably lobed lamina, leaf margins entire or serrated, venation palmate; inflorescence terminal, flowers 30–55 mm, six to nine petals, white or pink. Distribution: temperate eastern states of the USA, many countries in Southeast Asia including India and Japan [30,57].

Podophyllum hybrid—Habit: 60 cm high, peltate leaves, 10–40 cm, petiole green, almost centrally attached to the lamina, lamina palmatifid, two to five pointed lobes, at the base not lobed, light and dark green speckled, leaf margin entire. Petiole anatomy: vascular bundles arranged in a ring and central vascular bundles (number of vascular bundles: 15). Anatomy of the transition zone: fibre bundles branch out into smaller strands in the transition zone forming a net of fibre strands, merging into larger fibre bundles following the leaf veins. Additional data: see Table 2.

Crassulaceae J. Saint-Hilaire

Kalanchoe nyikae Engl.—Habit: perennial, erect or decumbent at the base, glabrous, glaucous plant, 60 cm to 2 m high; basal leaves almost circular, cuneate or slightly cordate at the base, margins entire, middle leaves peltate or auriculate, lamina 7–8 cm long, 6–7 cm wide, petioles 3–10 cm long, upper leaves lanceolate; inflorescences paniculate cymes up to 35 cm. Distribution: Kenya, Tanzania [30,58]. Petiole anatomy: several differently sized vascular bundles (number of vascular bundles: at least three). Anatomy of the transition zone: some fibre strands branch out; others extend into the lamina without branching. Additional data: see Table 2.

Umbilicus rupestris Dandy—Habit: geophyte, erect, glabrous, usually unbranched, 10–50 cm high; leaves crenate, roundish-peltate, 1.5–7 cm in diameter, central petiole insertion, petiole 4–25 cm long; inflorescence occupies 70%–90% of the stem length, racemose, flowers white, yellowish or reddish, often spotted, usually drooping. Distribution: Britain Islands, Mediterranean area, Arabian Peninsula. Ecology: temporarily humid locations, shady and humid growing places such as rocks and walls [18,30,58–60]. Petiole anatomy: smaller and larger vascular bundles arranged in one ring (number of vascular bundles: eight). Anatomy of the transition zone: vascular bundles from the petiole approach transition zone, run unbranched towards the lamina merging into larger bundles following the main veins. Additional data: see Table 2.

Eurphorbiaceae Jussieu

Macaranga tanarius (L.) Müll. Arg.—Habit: shrub to medium–high tree, evergreen; leaves alternate, simple, peltate, lamina circular to ovate, 8–32 cm × 5–28 cm, 9 radiating veins, margins entire, base rounded, apex acuminate, upper surface green, glabrous, lower surface paler, sometimes glabrous or occupied with sessile glands and simple hairs, petioles same as leaf length. Distribution: Southeast Asia and eastern Australia. Ecology: subtropical and tropical regions, common as a pioneer species in rainforest along the coast or cleared areas [30,61]. Petiole anatomy: vascular bundles arranged in two concentric rings and a central bundle (number of vascular bundles: 27), sclerenchyma associated with outer ring of vascular bundles, ring of sclerenchymatic tissue a few layers below epidermis, laticifers present. Anatomy of the transition zone: fibre strands from the petiole form a highly branched, dense network in the transition zone, then merge into larger bundles following the leaf veins into the lamina. Additional data: see Table 2.

Ricinus communis L.—This species is used in traditional medicine since the ancient Egyptian times. The oil of the seeds (castor oil, wonder oil) is used worldwide for a variety of medicinal purposes. Habit: shrubs or trees in tropical and subtropical regions up to 10 m high, in cooler regions often annual herbs 1–4 m in height; older stems brown or green, younger parts glaucous, annual plants often reddish or purplish; alternate leaves 30–50 cm, sometimes up to 100 cm large, papery, green, leaf shape palmately split with six to 11 ovate lobes, leaf margin serrated with unequal sized tips; inflorescence, up to 30 cm, fruit capsules dark red, seeds shiny brown, 8–11 mm; flowering and fruiting time throughout the year (tropical zones) or summer/late fall (temperate zones). Distribution: north-eastern

Africa, cultivated as an ornamental and a medicine plant throughout subtropical and temperate zones worldwide. Ecology: sunny, warm places, altitude 0–700 m, in some regions up to 2300 m, all kinds of habitats usually on rich soil, resistant to termites and drought [30,62–64]. Petiole anatomy: one ring of vascular bundles (number of vascular bundles: nine), ring of collenchymatic tissue below epidermis, central medullary canal, laticifers present (Figure 2F5). Anatomy of the transition zone: fibre strands from the petiole form a highly branched, dense network of fibre strands in the transition zone, then merging into larger bundles following the leaf veins into the lamina (Figure 2F4,F6)). Additional data: see Table 2.

Further studied species: *Jatropha podagrica* Hook.

Menispermaceae Jussieu

Perichasma laetificata Miers—Habit: climbing plant; leaves simple, petiole 9–12 cm long, hirsute, petiole insertion 4–4.5 cm from base of the lamina, lamina peltate, ovate, 17.5–22 cm long, 13–16 cm broad, leaf margin undulate, rounded at the base, bristly underneath, upper surface glabrous, palmately nerved, eight to nine nerves. Distribution: western Central Africa, from Nigeria to Angola [30,65]. Petiole anatomy: one ring of vascular bundles (number of vascular bundles: eight), ± closed ring of sclerenchyma fibres around vascular bundles. Anatomy of the transition zone: parallel fibre strands from petiole form a closed ring structure in transition zone, from this ring, single fibre bundles branch into leaf veins. Additional data: see Table 2.

Stephania delavayi Diels—Habit: slender herbaceous plant, vines, 1–2 m high, rarely branching; petiole and lamina mostly of similar length 3–7 cm long, leaf width ± equal to leaf length, leaf margins entire, lamina peltate, triangularly round, leaf base and tip blunt, both surfaces glabrous, abaxially pink–green and adaxially pale green to dark green, leaf thin papery, palmate, nine to 10-veined. Distribution: South and Central China and Myanmar. Ecology: in shrubland along fences and roadsides [30,40]. Petiole anatomy: one ring of vascular bundles (number of vascular bundles: eight), indistinct sclerenchyma fibres around vascular bundles, collenchymatic tissue below epidermis (Figure 2C5). Anatomy of the transition zone: parallel fibre strands from petiole form a closed ring structure in transition zone, from this ring, single fibre bundles branch into leaf veins (Figure 2C4,C6). Additional data: see Table 2.

Further studied species: *Stephania venosa* (Blume) Spreng.

Nelumbonaceae A. Richard

Nelumbo nucifera Gaertn.—The Indian lotus is cultivated as an ornamental plant and for its edible rhizomes and seeds. The seeds remain viable for several hundred years under certain conditions. Habit: perennial, rhizome forming, aquatic plant; petiole length 2 m and more, leaf stalks bare or papillose, very hard, leaves bluish green on the upper side, whitish green underneath, circular to oval shape, leaves entire, forming a slight funnel, 25–90 cm in diameter, very thin, water-repellent; pedicels of the flowers are longer than petioles, flowers pink or whitish yellow, 10–23 cm in diameter. Distribution: Southeast Asia, Southeast Europe, Australia, eastern USA. Ecology: temperate to tropical climates, different wetland habitats (ponds, lakes, lagoons, swamps and flood plains) with a depth of up to 2.5 m, altitude 400 m [30,38,57]. Petiole anatomy: scattered vascular bundles (number of vascular bundles: 63), large aerenchyma tubes, collenchyma and partly lignified tissue below epidermis (Figure 2G5). Anatomy of the transition zone: fibre bundles run parallel to the air channels into the transition zone, forming a symmetric net surrounding the cavities, eventually extending radially into the lamina (Figure 2G4,G6). Additional data: see Table 2.

Oxalidaceae R. Brown

Oxalis bowiei Aiton ex G. Don—also commonly known as Bowie's woodsorrel. Habit: bulbous perennial, stemless, pubescent; leaves basal, petiole 6–16 cm long, glandular pubescent, three leaflets, 3–6 cm long, obtuse, lobed, glandular pubescent, upper surface green, lower surface green to purplish; peduncles as long as leaves, inflorescences cymes, four to 12 flowers, flowers red, pink to deep rose pink. Distribution: Cape region of South Africa, introduced in California (USA), Korea and Western Australia. Ecology: disturbed areas, 300 m, occasionally grown horticulturally in mild temperate climates [30,63,66,67]. Petiole anatomy: central closed cylinder of vascular bundles (number of vascular bundles: seven) surrounded by thin layer of collenchyma, then parenchyma. Anatomy of the transition zone: cylinder of vascular bundles separate into three broad fibre strands merging into veins of the three leaflets. Additional data: see Table 2.

Passifloraceae Juss. ex Roussel

Passiflora coricaea Juss.—Habit: slender, climbing plant, perennial vine, 2–8 m or more, sparsely pilose on petioles, leaves, stems and stipules; petioles ca. 1–4 cm long, lamina elliptical, ca. 3–6 cm long, ca. 6–19 cm wide, with two or three lobes, petiole insertion point close to leaf margin, two lateral lobes 3–10 cm long, 2–7 cm wide, elliptical, pointed to weakly pointed, central lobe elliptical to obovate or present only as a blunt tip, leaf margins entire, three primary veins branching above the base; flowering stems (1–3 mm in diameter) greenish yellow to reddish purple, terete to somewhat compressed, base woody and cork-covered, flowers borne in leaf axils or in inflorescences, inflorescences 2.5–6.5 cm long, rarely up to 12 cm. Distribution: northern South America, Central America. Ecology: in shrubs and small trees in secondary successional areas, along edges of moist tropical forests, near rivers and streams, along the seashore, 0–1500 m [30,68]. Petiole anatomy: one central vascular bundle, five vascular bundles ± in a circle (three bundles in an arch, two bundles on the opposite side), layer of collenchymatic tissue below epidermis. Anatomy of the transition zone: fibre strands spread into lamina without branching. Additional data: see Table 2.

Saxifragaceae Jussieu

Darmera peltata (Torr. ex Benth.) Voss—Habit: herb with a perennial rhizome (1–5 cm in diameter); leaves emerging directly from rhizome, petiole 20–150 cm long, lamina 10–60 cm wide, sometimes up to 90 cm, peltate, centrally attached to petiole, leaf margins irregularly serrated, six to 15 lobed, upper side green, underside pale green, both sides ± glandular; flowering stems appear before leaves, erect, 30–100 cm high, sometimes up to 150 cm, inflorescences compound cymes, 30–150 cm large, 60 to 75 flowers, green to pinkish purple or white to rose, flowering time between April and July. Distribution: California and Oregon in the U.S., introduced in Czech Republic, France, Great Britain and Ireland. Ecology: between rocks in and around streams, altitude 30–1800 m [30,69]. Petiole anatomy: peripheral ring of vascular bundles and scattered central vascular bundles (number of vascular bundles: 24). Anatomy of the transition zone: fibre strands spread into the individual veins of the lamina without branching. Additional data: see Table 2.

Rodgersia podophylla A. Gray—Habit: perennial herb, 60–100 cm height; rhizome forming; stem glabrous; petiole 15–30 cm long, pilose, leaves palmately compound, five leaflets, obovate, 15–30 cm long, 10–25 cm wide, lobed apex (three to five lobes), adaxially glabrous, abaxially pilose along the veins, serrated margin, apex acuminate, cauline leaves smaller than basal leaves; many flowered inflorescence, sepals white, petals absent, flowering time June to July. Distribution: Japan, Korea, China, introduced into Czechoslovakia, Great Britain, Norway. Ecology: shaded slopes [30,70]. Petiole anatomy: vascular bundles arranged in several rings (number of vascular bundles: 63), sclerenchymatic bundle caps visible, sclerenchymatic caps surrounding outer ring of vascular bundles form an almost closed ring. Anatomy of the transition zone: fibre strands from the petiole run mostly unbranched, only few branching bundles, through the transition zone into leaflets. Additional data: see Table 2.

Tropaeolaceae Juss. ex de Candolle

Tropaeolum tuberosum Ruiz. and Pav.—This species, also known as "Mashua", is used as a tuber crop plant in the cool–temperate Andes. Habit: herbaceous perennial, > 2 m high, 1 m in diameter; tuber white to yellow, occasionally red, flesh yellow; leaves peltate, three to five lobed, 4–6 cm long, 5–7 cm broad, palmate venation petioles glabrous, reddish, 4–20 cm long, plant uses long leaf stalks as tactile petioles for climbing; flowers long-stalked, solitary, variable in colour such as orange, yellow or scarlet. Distribution: higher Andes of Peru, Colombia, Ecuador and Bolivia, cultivated in parts of northern Argentina and Chile, also introduced in New Zealand. Ecology: altitude 2400–4300 m, annual mean temperatures 8–11°C [71–73]. Petiole anatomy: ring consisting of seven vascular bundles, distinct layers of collenchyma underneath the epidermis (Figure 2D5). Anatomy of the transition zone: fibre strands unite into dense, knot-like structure, individual fibre strands extend from there into leaf veins (Figure 2D4,D6). Additional data: see Table 2.

Urticaceae Jussieu

Cecropia peltata L.—Habit: tree, 15–25 m; leaves 10–60 cm in length and width, palmately divided into seven to 11 lobes, upper surface scaly, lower side hairy, petioles 20–50 cm long. Distribution: Mexico to northern South America, introduced in Africa and Asia. Ecology: wet to deciduous forests, as secondary vegetation, in pastures and roadsides, altitude 0–1800 m [30,74]. Petiole anatomy: peripheral ring of vascular bundles (number of vascular bundles: 23), layers of collenchyma below epidermis. Anatomy of the transition zone: fibre strands from the petiole form a highly branched, dense network, merging into larger bundles following the leaf veins into the lamina. Additional data: see Table 2.

Pilea peperomioides Diels—Habit: herbaceous perennial plant, erect or ascending, wild plants up to 23 cm, in cultivation up to 60 cm in height; robust stems, greenish to dark brownish; leaves spirally arranged, lamina pale green underneath, upper side green, somewhat succulent, three veins, three to four lateral veins on each side, petioles 2–17 cm long, leaves peltate, elliptic, 4–7 cm long and wide, base rounded, apex round or obtuse; inflorescence solitary, 18–28 cm, flowers pinkish cream, in clusters. Distribution: China (West Yunnan and Southwest Sichuan), introduced in Belgium. Ecology: on shady, damp rocks in forests or cliff-ledges, on humus-covered boulders, altitude 1500–3000 m, very rare in the wild, possibly endangered in its native habitat [30,75,76]. Petiole anatomy: few centrally arranged vascular bundles surrounded by parenchymatic tissue (number of vascular bundles: six). Anatomy of the transition zone: fibre strands spread out in transition zone, proceed unbranched into lamina. Additional data: see Table 2.

3.2.5. Polypodiopsida
Marsileaceae Mirbel

Marsilea strigosa Willd.—Habit: aquatic, heterosporous fern with long stolons, thick creeping subterranean rhizome; leaves with four green, pubescent leaflets, 0.8–1.5 cm in diameter, petioles slender, 10–15 cm long, two types of leaf variants: (1) floating leaves, glabrous, with long slender petiole, (2) free standing leaves, rigid petiole, tomentose blade; reproduces clonally and sexually, producing stoloniferous structures and sporocarps, brown, sessile, spherical sporocarps (0.5–0.3 cm diameter) at the base of petioles. Distribution: Afghanistan, Algeria, Baleares, eastern European Russia, Egypt, France, Italy, Kazakhstan, Morocco, Portugal. Ecology: short life cycle, Mediterranean temporary pools, among amphibious vegetation on various non-calcareous substrates [30,77]. Petiole anatomy: vascular tissue in central cylinder. Anatomy of the transition zone: cylinder separates into several fibre strands that run into the leaflets. Additional data: see Table 2.

3.3. Classification of Strengthening Structures in the Petiole–Lamina Transition Zone

The microscopic examination of peltate leaves from different plant species revealed different types of transition zones formed by fibre strands at the joint between the lamina and petiole. The types vary from virtually absent or very simple to very complex structures, with gradual transitions in intensity as well as in the form and shape of the branching system. In total, seven different types with one subtype have been identified (including a simple unbranched type).

1. Unbranched type (A): The fibre strands in the petiole continue unbranched through the petiole–lamina junction into the lamina (Figure 2A). The strands follow the leaf veins.
2. Simple branching type (B): Some fibre strands in the petiole branch in the transition zone and then extend into the leaf veins while others proceed unbranched into the lamina (Figure 2B). The fibre strands branching in the transition zone can either run in into the same leaf vein of the lamina or follow a different direction into different leaf veins, forming a connection between, e.g., opposite sides of the lamina. All species with at least one visible branching fibre strand were assigned to this category.
3. Ring-like structure (Type C): The fibre strands from the petiole run parallel towards the petiole–lamina junction, then merge into a fibre ring without any visible branching. This ring is oriented parallel to the lamina and connects all the strands from the petiole with each other. Starting from the ring, the strands spread into the leaf veins (Figure 2C). This structure was found only in *Stephania* and *Perichasma*.
4. Knot-like structure (Type D): Isolated fibre strands from the petiole assemble to a dense knot-like structure in the transition zone. It is not possible to distinguish between individual fibre strands in this knot. A few stronger fibre strands spread out from these structures and continue into the lamina (Figure 2D).
5. Net-like structure (Type E and F): The individual fibre strands the petiole form a net-like structure in the transition zone. Individual fibre strands are still visible. Fibres branch and merge in different intensities, forming nets of different densities. Two subtypes were identified here: (1) slight to moderate branching (Figure 2E) and (2) intensive branching (Figure 2F) of the fibre strands. From this net, several fibre bundles spread out into the lamina following the larger veins.
6. Radially branching type (water plants, Type G): The fibre strands from the petiole are united in a dense structure with large fibre strands and big intercellular cavities for gas exchange. The fibre strands run parallel to the air channels branch in the transition zone and extend into the lamina in a radial pattern (Figure 2G).
7. Diffuse branching type (water plants, Type H): Parallel fibre strands from the petiole branch in the transition zone forming a diffuse net of fibres of different sizes (Figure 2H). Air channels are visible.

The unbranched and simple branching types are quite common among the analysed species. Eight species found in six orders throughout the plant kingdom, from magnoliids and monocots to eudicots, show no conspicuous strengthening structure in the joint of the petiole and lamina (Figure 4). The simple branching type occurs in twelve species and six orders with the Araceae (monocots) and Piperaceae (magnoliids) containing the majority of the twelve taxa (Araceae: five species, Piperaceae: three species). All five species from Araceae (five different genera) show this type of branching. The net-like structure could be found in five orders. While the slight to moderate branching intensity is found in Ranunculales (*Podophyllum*) and two core eudicot orders (Cucurbitales and Apiales, four species), the intensive branching is only present in two rosid orders (Malpighiales and Rosales, four species). The knot-like structure occurs in two orders from the core eudicots (Brassicales and Apiales, one species each). The ring-structure was only found in the Ranunculales, in the Menispermaceae. All three investigated species from the Menispermaceae show this strengthening structure in the petiole–lamina transition zone. The two branching types in water plants were found in Nymphaeales (four species). The radial branching type additionally appears in the Proteales (Nelumbonaceae, *Nelumbo nucifera*).

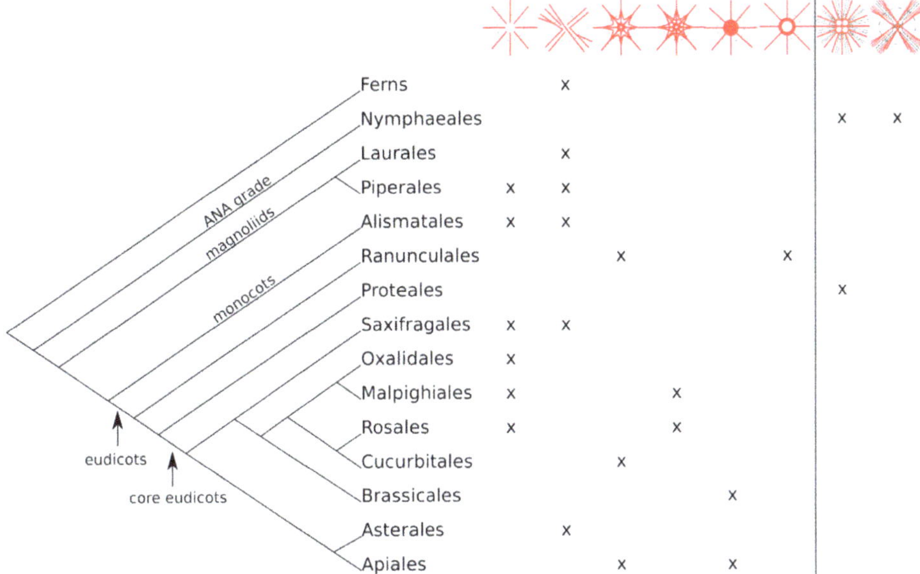

Figure 4. Distribution of the different strengthening structures in the petiole–lamina junction of peltate leaves within the plant kingdom. Strengthening structures (from left to right): unbranched, simple branching, net-like (intensities 1 and 2), knot-like, ring-like, radial (water plant), diffuse (water plant). The unbranched and simple branching type are the most common types found in six orders, followed by the net-like structure. All other structures only occur in one to two orders.

When comparing the proportion of lignified tissue in the petiole of the different types of strengthening structures, the unbranched, simple branching and water plant diffuse branching type seem to have lower amounts than the net-like structure, intensity 2 and the ring-like structure (Figure 3A). The water content in petiole, venation and intercostal areas shows a high variance in most categories. However, the net-like structure, intensity 2 and partly ring-like structure and knot-like structure seem to show lower water content than the other categories (Figure 3B–D).

4. Discussion

4.1. The Peltate Leaf Shape and Its Representatives in the Plant Kingdom

Since the lists published by de Candolle (1899), Troll (1932) and Ebel (1998), the distribution of the peltate leaf shape in the plant kingdom has not been focus of many studies [18,19,24]. This study aimed to provide an up to date, comprehensive list of peltate plant species including information about morphology, anatomy, distribution, and habitat. The peltate leaf shape was found in all clades of angiosperms including basal angiosperms in the ANA grade (Table 1). The order Nymphaeales with the Nymphaeaceae and Cabombaceae contains peltate aquatic plants differing in shape and size, but the leaves, without exception, are floating. While most of the peltate leaves of the genus *Nelumbo* in the Nelumbonaceae are raised above the water surface, *Nelumbo* also forms floating leaves [38,57]. Ebel (1998) noted that, in these two groups, the ratio of peltate to non-peltate genera is shifted in favour of the peltate leaved type [18]. In the entire clade of magnoliids, only two families with peltate species have been identified (Piperaceae, Hernandiaceae). In the Piperaceae, especially in the species-rich genus *Peperomia* with about 1600 species [78], many peltate representatives could be found with a marginal to central petiole insertion point (Table 1). As Ebel (1998) already observed [18], the *Peperomia* has its main distribution area in the Neotropics.

Peltate leaves are unusual among monocotyledons, being virtually limited to the Araceae [18]. Ten genera with peltate leaved species could be determined, which is more when compared to the other investigated families. As land plants, they are mostly found in subtropical areas (Table 1). Ebel (1998) suggests that the peltation in Araceae is likely related to the leaf venation, which is pinnate or palmate instead of the typical parallel venation of monocots [18,79,80] and observed in all peltate leaved Araceae. *Arisaema* and *Amorphophallus* occupy a special position, since both are not only centrally peltate but also form pinnate leaves, while the remaining genera show an eccentric peltation and have a simple entire lamina.

Peltation is most widespread in eudicots (Table 1). However, the occurrence is not limited to a few orders or families, as in monocots and magnoliids. In each order, one to four families with up to nine genera contain species with peltate leaves (e.g., Euphorbiaceae, Asteraceae, Ranunculaceae, Gesneriaceae, Menispermaceae, Table 1).

Among the studied species, leaves with the eccentric peltation are more common compared to those with a central attachment of the petiole (Table 1). Genera rich in centrally peltate species (e.g., *Hydrocotyle*, *Umbilicus*, *Oxalis*, *Podophyllum* and several genera of the Saxifragaceae) are native to meridional and temperate floral zones (or the southern hemispheric equivalent). In the neotropical and large parts of the palaeotropical floral zones, centrally peltate leaves are not as common as eccentric peltate leaves, indicating a connection between central peltation and geographic distribution, but the data does not give any explanation for this. Species with central peltation in the tropical and subtropical zones are mainly restricted to Nymphaeaceae, Oxalidaceae and Araliaceae. For Nymphaeaceae, the central peltation and the circular shape of the leaves can be explained as an adaptation to floating on the water surface, but this does not explain the peltate leaves in terrestrial plants.

The great majority of species with peltate leaves are perennial herbs. Already, Ebel (1998) noted that most of the taxa with peltate leaves are rhizomatous, tuberous or stoloniferous perennials growing in wet, humid or alternately wet habitats [18]. While Troll (1932) rejected a correlation between peltation and ecology [19], Friedrich Ebel (1998) pointed towards a connection between peltation and certain chorological and habitat conditions [18]. Data from this study suggest that there may be relations between the peltate leaf shape and species distribution, habitat or phylogeny supporting Ebel's findings. Nevertheless, the observations at this point can only be regarded as indications.

4.2. Supporting a Peltate Lamina—Different Adaptations in the Petiole–Lamina Junction

The anatomical analysis of the petiole–lamina transition of 41 peltate leaved species from 18 families revealed seven different types of strengthening structures.

This study focused on reinforcing elements consisting of xylem elements and sclerenchymatic fibres, which are mainly found as vascular bundle sheaths or sclerenchyma caps associated with the vascular bundles [2,3,17]. However, in addition to these fibres, other strengthening tissue, such as collenchyma but also epidermal cell walls or the turgor pressure of living tissues, especially the parenchyma contribute to mechanical stability of plants [1,14,15,81]. As measurements of strengthening tissue were conducted only on a small number of samples and for some categories only a small number of leaves were available, the conclusions made in the discussion should be seen as tendencies.

4.2.1. Simple Structures in the Petiole–Lamina Junction—The Unbranched and Simple Branching Structure

The unbranched structure, Type A, is a most common type (eight species, Table 2) among the examined peltate leaves. Since the leaves with unbranched fibre strands (Figure 2A) at first glance seem to have developed a less effective strengthening structure in the transition zone from the petiole to lamina, a closer look at *Amorphophallus konjac* points towards an explanation how, in this case, a very large lamina with up to 2 m in diameter [44] is stabilised: a combination of a large number of vascular bundles (180) with lignified xylem, on the one hand, and layers of collenchyma tissue associated with the outer vascular bundles on the other hand. These layers of collenchyma are a common mechanical tissue in the Araceae [15]. Another possibility to hold up and stabilise the peltate lamina can be observed in *Umbilicus rupestris* and *Pilea peperomioides*, both forming more or less succulent leaves and growing in shady, damp habitats [58,76]. Both species show high values in water loss (over 95%) displaying the high water content of the succulent leaves, low to medium amount of sclerenchymatic strengthening tissue in the petiole (up to 10.7%, Table 2) and no collenchyma. This suggests that the stability of the lamina of both species is mainly ensured via the turgor pressure in the parenchymatic tissues.

The simple branching structure (Type B) is characterized by the branching of some fibre strands while others run into the lamina unbranched. In its simplicity, it is a common structure in the transition zone from the petiole to lamina with 12 species from six orders sorted into this category (Figure 4). The families and genera sorted in this category show different anatomical features, but all show this type of branching. Certainly, further subcategories could be defined with more in-depth research. With eight of the 12 taxa sorted into this category, the majority of the simple branched species can be found in the more basal orders of the angiosperms.

Characteristic for this group is the relatively low proportion of strengthening tissue (1–10%, Table 2 and Figure 3A) suggesting that these species, similar to the unbranched species, need other additional strategies to support the peltate lamina.

Most of the studied Araceae exhibit the simple branching type. All of them show high numbers of scattered vascular bundles in the petiole (30–90, Table 2) and associated layers of collenchyma tissue (common in Araceae [15]), which stabilise the partly very large leaves with small ratios of petiole area to lamina area (0.04%–0.07%). The medium to high water content (80–95%) indicate that these species also strongly depend on turgor pressure to support the peltate laminae. In the petiole–lamina junction, the cross-linking of opposite sides of the lamina could be observed. Some of the fibres branch with individual arms extending into different parts of the lamina (Figure 2B), providing extra stability.

All studied *Peperomia* species (Piperaceae) display the simple branching type as well. In contrast to Araceae, the *Peperomia* species show significantly lower numbers of vascular bundles in the petiole arranged in a circle (6–8, Table 2). The studied neotropical *Peperomia* species are very low growing herbaceous perennials with small leaf blades [30], display high ratios of petiole to lamina area (0.25% for *P. sodiroi*, Table 2) and a high water content (about 95%). Given the growth conditions, habit and the large petiole to leaf ratio, *Peperomia* obviously does not need a complex strengthening structure in the petiole–lamina junction but supports most of its own weight by turgor pressure.

4.2.2. The Net-Like Branching Structure

The net-like structures (Types E and F) could be observed in several species from different genera and families. Due to the different intensities of branching and interconnecting of the fibre strands, this category is divided into two different subtypes, based on the same reticulate structure.

The slightly branched reticulate structure (Type E) is mainly found in *Begonia*. This mega-diverse genus exceeding 1800 accepted species (over 2000 species estimated) and a pantropical distribution shows a variety of growth forms including herbs, shrubs, acaulescent species, succulent or woody stems or lianoid species [51,82]. The leaves of *Begonia* show a great diversity as well as with several sections of the genus containing peltate species [51]. Although differing in shape, size and texture, the leaves of Begonia reveal a similar structure in the petiole cross sections with collenchyma underneath the epidermis and the lignified xylem elements as additional strengthening elements. The branching and crosslinking of the xylem fibre strands in the petiole–lamina junction ensure the stabilisation of the peltate lamina (Figure 2E). The low proportions of strengthening tissue (4%–8%, Table 2) and high water content of the leaf (90%–95%) again indicate that these species strongly depend on turgor pressure to support their laminae with collenchyma tissue as counterpart.

Leaves with strongly branched fibre strands forming a reticulate structure (Type F) are mainly found in shrubs and trees form Euphorbiaceae such as *Jatropha podagrica*, *Macaranga tanarius* and *Ricinus communis* (Table 2). Despite morphological differences, e.g., leaf shape and origin from three different continents, anatomical similarities are apparent in the peripheral circle of vascular bundles with distinct xylem tissue and layers of mechanical tissue (collenchyma or sclerenchyma) associated with the vascular bundles. *Macaranga tanarius* shows a closed ring of sclerenchyma tissue associated with the outer vascular bundles, apparent in the highest proportion of strengthening tissue (18.8%) when compared to *R. communis* (13.4%) and *J. podagrica* (10.8%) (Table 2 and Figure 3A). In comparison to the low intensity of net-like branching in *Begonia*, the intensive crosslinking in Euphorbiaceae, becomes apparent in the higher ratio of strengthening tissue (Figure 3A). However, both subtypes show more or less similar ratios of petiole area to lamina area (Table 2). While the mostly herbaceous *Begonia* grow in the ground layers of the forest, the three Euphorbiaceae species grow into shrubs or trees and thus, the leaves are exposed to higher external forces such as wind and rainfall. Here, the high proportion of strengthening tissue and crosslinking might provide the needed stability. The fact that the species share the same type of transition structure supports the assumption that the structure is typical for Euphorbiaceae, rather than influenced by origin or environmental conditions.

4.2.3. Dense Structures in the Petiole–Lamina Junction—The Ring- and Knot-Like Structures

The peculiar ring-like structure in the petiole–lamina transition (Type C) was observed only in Menispermaceae (*Perichasma laetificata* and *Stephania delavayi*, Figure 2C). Both show high proportions of sclerenchyma tissue in the petiole (*P. laetificata*: 18.77%, *S. delavayi*: 12.14%, Table 2 and Figure 3A). Both species are climbers but are native to different continents [30,40,65]. The complex ring-like structure in the transition zone and the high proportion of sclerenchyma probably is a specific for one family, as in the Euphorbiaceae.

The fibre-dense knot-like structure (Type D) was found in two species from two different orders, *Tropaeolum tuberosum* (Brassicales) and *Hydrocotyle vulgaris* (Apiales) (Figure 4). In both species the fibre strands merge into a dense knot in the transition zone. As a climbing plant, *T. tuberosum* is found in the Andean regions of South America [71]. *Tropaeolum tuberosum* is climbing with leaf-stalk tendrils [71] i.e., the petioles carry most of the plant's weight. High amounts of strengthening tissue in the petiole (12.8%, Table 2) most probably ensure a secure connection between petiole and support. The anatomical analysis with light microscopy was not able to differentiate between individual fibre strands in the junction. Further analysis of such structures in other peltate species of *Tropaeolum* and *Hydrocotyle*, applying CT- or MRI-scanning, like in Sacher et al. (2019) will shed further light on the fibre orientation in this type of joint [13].

4.2.4. Specifics of Water Plants

Aquatic plants with floating leaves are treated separately here, because of the specific growth under which it is generally not necessary to support the weight of the leaves (except *Nelumbo*, some *Nymphaea* species or *Marsilea*).

The analysis of the transition zone of several peltate leaves of water plants revealed two different types, (1) the radial branching type (Type G, *Nelumbo nucifera*, *Nymphaea colorata*) and (2) the diffuse branching type (Type H, *Victoria cruziana*, *V. amazonica*, *Euryale ferox*). Though plants with floating leaves are not exposed to wind or need the strengthening tissue to raise the lamina above the water, they nonetheless need to withstand other challenges such as resisting water turbulences, flooding, drought or ensuring gas exchange throughout the plant body. All investigated aquatic species show the characteristically large ventilation channels in the petioles (Figure 2G,H), which are slightly different in each species. All species have scattered vascular bundles, except for some larger ones that are located between the ventilation channels. In addition, some layers of collenchyma tissue have been found underneath the epidermis in each species. Compared to other peltate leaved species, *E. ferox* has a remarkably high ratio of petiole to lamina area (being

0.28%, Table 2). A wider petiole could be advantageous for the gas exchange between lamina and the submersed rhizome. In addition, a robust petiole securely connected to the lamina could prevent the lamina from tearing off due to water movement or other mechanical stresses.

The genus *Nelumbo* holds a special position among the analysed aquatic plants with floating leaves. *Nelumbo* produces floating leaves comparable to the other aquatic species but mostly leaves that are raised above the water [38,57]. The strengthening structure in the petiole–lamina junction is similar to the one of *Nymphaea*. The large amount of strengthening tissue in the petiole (19.3%, Table 2), when compared to *N. colorata* with floating leaves (5,7%, Table 2) indicates that higher amounts of strengthening tissue are needed to withstand the higher gravitational forces and external stresses such as wind and rain that are affecting the emergent lamina.

Cabomba aquatica could not be assigned to one of the previously described types. *Cabomba* predominantly forms submerged leaves while floating leaves appear only at flowering time [83]. The anatomy of the petiole, including large ventilation channels, is similar to the petioles of other aquatic plants. However, the central position of just two fibre strands is unique. The strands spread into an undefined dense structure in the transition zone and subsequently divide into four strands that finally unevenly branch within the lamina. With this central position of vascular bundles and fibres, further analysis is needed to clarify the detailed structure of the transitions zone of *C. aquatica*.

4.2.5. Conclusions and Outlook

Since the analysis of peltate species from 18 different families revealed seven distinct types of structures in the petiole–lamina junction, it is probable to find more principles of strengthening structures in the remaining 22 families with peltate leaved plant species. Future studies will follow up on this hypothesis.

When comparing water content, amount of strengthening tissue and anatomical observations for the seven different principles, especially the dense structures (knot-like, ring-like), structures with intense branching (net-like, intensity 2) and with large numbers of vascular strands (simple branching) are likely to show a higher stiffness and rigidity than those structures with less strengthening tissue and mostly depending on turgor pressure. Biomechanical analyses of the different types will clarify this theoretical assumption and are already in progress.

Although the anatomical analysis, applying light microscopy, provided good results to obtain an insight into the different structures in the petiole–lamina transition zone, other methods are needed to allow for a more detailed analysis of the orientation, branching, and merging of the fibre strands. These investigations, i.e., CT (computed tomography)- and MRI (magnetic resonance imaging)-scanning are currently under way and will be combined with the biomechanical data of the different structural elements, eventually resulting in models that will be used for bioinspired structures in lightweight constructions such as carbon-concrete components.

Supplementary Materials: The following are available online at https://www.mdpi.com/2313-7673/6/2/25/s1, Document S1: Profiles of further studied species, Table S1: List of peltate species among vascular plants. Figure S1: Leaf morphology and anatomy of *Cabomba aquatica* Aubl., Figure S2: Leaf morphology and anatomy of *Nymphaea colorata* Peter, Figure S3: Leaf morphology and anatomy of *Victoria amazonica* (Poepp.) J.C. Sowerby, Figure S4: Leaf morphology and anatomy of *Peperomia argyreia* (Miq.) E. Morren, Figure S5: Leaf morphology and anatomy of *Caladium* hybrid, Figure S6: Leaf morphology and anatomy of *Remusatia vivipara* (Roxb.) Schott, Figure S7: Leaf morphology and anatomy of *Colocasia esculenta* (L.) Schott, Figure S8: Leaf morphology and anatomy of *Jatropha podagrica* Hook., Figure S9: Leaf morphology and anatomy of *Stephania venosa* (Blume) Spreng.

Author Contributions: Conceptualization, T.L. and C.N.; investigation, J.W.; methodology, J.W.; writing—original draft preparation, J.W. and A.R.; writing—review and editing, T.L. and C.N.; supervision, T.L. and C.N. All authors have read and agreed to the published version of the manuscript.

Funding: This research was funded by the Deutsche Forschungsgemeinschaft (DFG, German Research Foundation)—SFB/TRR 280. Project-ID: 417002380.

Institutional Review Board Statement: Not applicable.

Informed Consent Statement: Not applicable.

Data Availability Statement: Data is contained within the article or Supplementary Materials.

Acknowledgments: The authors thank the staff of the Botanical Garden Dresden, namely Barbara Ditsch and Philipp Weiner and the staff of the Ecological–Botanical Garden Bayreuth, namely Marianne Lauerer for providing the plant material. Open Access Funding by the Publication Fund of the TU Dresden.

Conflicts of Interest: The authors declare no conflict of interest.

References

1. Faisal, T.R.; Abad, E.M.K.; Hristozov, N.; Pasini, D. The impact of tissue morphology, cross-section and turgor pressure on the mechanical properties of the leaf petiole in plants. *J. Bionic Eng.* **2010**, *7*, S11–S23. [CrossRef]
2. Adams, W.W., III; Terashima, I. *The Leaf: A Platform for Performing Photosynthesis*; Springer: Berlin, Germany, 2018.
3. Sitte, P.; Ziegler, H.; Ehrendorfer, F.; Bresinsky, A. *Strasburger Lehrbuch Der Botanik*, 34th ed.; Gustav Fischer Verlag: Stuttgart/Jena/Lübeck/Ulm, Germany, 1998.
4. Raven, P.H.; Evert, R.F.; Eichhorn, S.E. *Biologie Der Pflanzen*, 3rd ed.; Walter de Gruyter: Berlin, Germany; New York, NY, USA, 2000.
5. Vogel, S. Drag and reconfiguration of broad leaves in high winds. *J. Exp. Bot.* **1989**, *40*, 941–948. [CrossRef]
6. Niklas, K.J. Petiole mechanics, light interception by lamina, and "Economy in Design". *Oecologia* **1992**, *90*, 518–526. [CrossRef] [PubMed]
7. Raupach, M.R.; Thom, A.S. Turbulence in and above plant canopies. *Annu. Rev. Fluid Mech.* **1981**, *13*, 97–129. [CrossRef]
8. Gosselin, F.P.; De Langre, E. Drag reduction by reconfiguration of a poroelastic system. *J. Fluids Struct.* **2011**, *27*, 1111–1123. [CrossRef]
9. Niklas, K.J. A mechanical perspective on foliage leaf form and function. *New Phytol.* **1999**, *143*, 19–31. [CrossRef]
10. Niklas, K.J. *Plant Biomechanics: An Engineering Approach to Plant Form and Function*; University of Chicago Press: Chicago, IL, USA, 1992.
11. Levionnois, S.; Coste, S.; Nicolini, E.; Stahl, C.; Morel, H.; Heuret, P. Scaling of petiole anatomies, mechanics and vasculatures with leaf size in the widespread neotropical pioneer tree species cecropia obtusa trécul (*Urticaceae*). *Tree Physiol.* **2020**, *40*, 245–258. [CrossRef]
12. Niklas, K.J. Flexural Stiffness allometries of angiosperm and fern petioles and rachises: Evidence for biomechanical convergence. *Evolution* **1991**, *45*, 734–750. [CrossRef] [PubMed]
13. Sacher, M.; Lautenschläger, T.; Kempe, A.; Neinhuis, C. Umbrella leaves—Biomechanics of transition zone from lamina to petiole of peltate leaves. *Bioinspir. Biomim.* **2019**, *14*, 046011. [CrossRef]
14. Leroux, O. Collenchyma: A versatile mechanical tissue with dynamic cell walls. *Ann. Bot.* **2012**, *110*, 1083–1098. [CrossRef]
15. Keating, R.C. Collenchyma in *Araceae*: Trends and relation to classification. *Bot. J. Linn. Soc.* **2000**, *134*, 203–214. [CrossRef]
16. Niinemets, Ü.; Fleck, S. Petiole mechanics, leaf inclination, morphology, and investment in support in relation to light availability in the canopy of liriodendron tulipifera. *Oecologia* **2002**, *132*, 21–33. [CrossRef] [PubMed]
17. Evert, R.F. *Esau's Plant Anatomy: Meristems, Cells, and Tissues of the Plant Body: Their Structure, Function, and Development*; John Wiley & Sons: Hoboken, NJ, USA, 2006.
18. Ebel, F. Die schildblättrigkeit krautiger angiospermen-sippen in ihrer beziehung zu standort und verbreitung. *Flora* **1998**, *193*, 203–224. [CrossRef]
19. Troll, W. Morphologie der schildförmigen blätter. *Planta* **1932**, *17*, 153–230. [CrossRef]
20. Troll, W. *Vergleichende Morphologie Der Höheren Pflanzen, Vegetationsorgane*; Gebrüder Bornträger: Berlin, Germany, 1939; Volume 1.
21. Troll, W.; Meyer, H.-J. Entwicklungsgeschichtliche untersuchungen über das zustandekommen unifazialer blattstrukturen. *Planta* **1955**, *46*, 286–360. [CrossRef]
22. Roth, I. Beiträge zur entwicklungsgeschichte der schildblätter. *Planta* **1952**, *40*, 350–376. [CrossRef]
23. Givnish, T.J.; Vermeij, G.J. Sizes and shapes of liane leaves. *Am. Nat.* **1976**, *110*, 743–778. [CrossRef]
24. De Candolle, C. Sur les feuilles peltées. *Bull. Trav. Soc. Bot. Genève* **1899**, *9*, 1–51.
25. Uittien, H. Ueber den zusammenhang zwischen blattnervatur und sprossverzweigung. *Recl. Trav. Bot. Néerl.* **1929**, *25*, 390–483.
26. Chase, M.W.; Christenhusz, M.J.M.; Fay, M.F.; Byng, J.W.; Judd, W.S.; Soltis, D.E.; Mabberley, D.J.; Sennikov, A.N.; Soltis, P.S.; Stevens, P.F. An update of the angiosperm phylogeny group classification for the orders and families of flowering plants: APG IV. *Bot. J. Linn. Soc.* **2016**, *181*, 1–20.
27. Roth, I. Zur entwicklungsgeschichte des blattes, mit besonderer berücksichtigung von stipular und ligularbildungen. *Planta* **1949**, *37*, 299–336. [CrossRef]
28. Leinfellner, W. Über die peltaten kronblätter der sapindaceen. *Österr. Bot. Z.* **1958**, *105*, 443–514. [CrossRef]
29. Missouri Botanical Garden EFloras. Available online: http://www.efloras.org (accessed on 29 September 2020).

30. Royal Botanic Garden, Kew POWO, Plants of the World Online. Available online: http://www.plantsoftheworldonline.org/ (accessed on 29 September 2020).
31. GBIF Secretariat GBIF.Org (Global Biodiversity Information Facility). Available online: http://www.gbif.org (accessed on 29 September 2020).
32. Missouri Botanical Garden Tropicos.Org. Available online: http://www.tropicos.org/ (accessed on 29 September 2020).
33. The Plant List, Version 1.1. Available online: http://www.theplantlist.org/ (accessed on 29 September 2020).
34. Cole, T.C.; Hilger, H.H. *Angiosperm Phylogeny Poster–Flowering Plant Systematics*; Freie Universität: Berlin, Germany; Missouri Botanical Garden: St. Louis, MO, USA, 2019.
35. Ellis, B.; Daly, D.C.; Hickey, L.J.; Mitchell, J.D. *Manual of Leaf Architecture: Morphological Description and Categorization of Dicotyledonous and Net-Veined Monocotyledonous Angiosperms*; Leaf Architecture Working Group: Washington, DC, USA, 1999.
36. Jäger, E. Die Pflanzengeographische ozeanitätsgliederung der holarktis und die ozeanitätsbindung der *Pflanzenareale*. *Feddes Repert.* **1968**, *79*, 157–335. [CrossRef]
37. Pott, R. *Allgemeine Geobotanik: Biogeosysteme Und Biodiversität*; Springer: Berlin, Germany, 2014.
38. Wu, Z.; Raven, P.H.; Chen, J. *Flora of China: Caryophyllaceae through Lardizabalaceae*; Science Press: Beijing, China, 2001; Volume 6.
39. Jain, A.; Singh, H.B.; Kanjilal, P.B. Economics of Foxnut (Euryale Ferox Salisb.) Cultivation: A Case Study from Manipur in North Eastern India. *Indian J. Nat. Prod. Resour.* **2010**, *1*, 63–67.
40. Wu, Z.; Raven, P.H.; Hong, D. *Flora of China: Menispermaceae through Capparaceae*; Science Press: Beijing, China, 2008; Volume 7.
41. De Candolle, C. Piperaceae sodiroanae. In *Bulletin de l'Herbier Boissier*; Imprimerie Romet: Geneva, Switzerland, 1898; Volume 6, p. 506.
42. Trelease, W. *The Piperaceae of Northern South America*; University of Illinois Press: Urbana, IL, USA, 1950.
43. Tebbs, M. Revision of piper (*Piperaceae*) in the new world 3. The taxonomy of piper sections lepianthes and radula. *Bull. Br. Mus. Nat. Hist. Bot.* **1993**, *23*, 1–50.
44. Wu, Z.; Raven, P.H.; Hong, D. *Flora of China: Acoraceae through Cyperaceae*; Science Press: Beijing, China; Missouri Botanical Garden Press: St. Louis, MO, USA, 1994; Volume 23.
45. Ebel, F.; Mühlberg, H.; Hagen, A. Ökomorphologische Studien an Schwingmoorpflanzen. *Hercynia-Ökol. Umw. Mitteleur.* **1989**, *26*, 432–444.
46. Jäger, E. *Rothmaler-Exkursionsflora von Deutschland. Gefäßpflanzen: Grundband*; Springer: Berlin/Heidelberg, Germany, 2017.
47. Lansdown, R.V. Hydrocotyle Vulgaris. The IUCN Red List of Threatened Species 2014: E.T164201A42415437. Available online: https://www.iucnredlist.org/species/164201/42415437 (accessed on 8 October 2020).
48. Wu, Z.; Raven, P.H.; Hong, D. *Flora of China: Clusiaceae through Araliaceae*; Science Press: Beijing, China, 2007; Volume 13.
49. Missouri Botanical Garden Gardnening Help, Plant Finder: Syneilesis Palmata. Available online: https://www.missouribotanicalgarden.org/PlantFinder/PlantFinderDetails.aspx?taxonid=277696&isprofile=0& (accessed on 27 November 2020).
50. Wu, Z.; Raven, P.H.; Hong, D. *Flora of China: Asteraceae*; Science Press: Beijing, China; Missouri Botanical Garden Press: St. Louis, MO, USA, 2011; Volume 20.
51. Doorenbos, J.; Sosef, M.S.M.; De Wilde, J.J.F.E. *The Sections of Begonia Including Descriptions, Keys and Species Lists (Studies in Begoniaceae VI)*; Wageningen Agricultural University: Wageningen, The Netherlands, 1998.
52. Otto, C.F.; Dietrich, A.G. Begonia peltata. *Allg. Gartenzeitung* **1841**, *9*, 58–59.
53. Twyford, A. *The Evolution of Begonia Section Gireoudia in Central America*; University of Edinburgh, Royal Botanic Gardens Edinburgh: Edinburgh, UK, 2000.
54. Bailey, L.H. *Manual of Cultivated Plants / Most Commonly Grown in the Continental United States and Canada*; MacMillan: New York, NY, USA, 1954.
55. Burt-Utley, K. *A Revision of Central American Species of Begonia Section Gireoudia (Begoniaceae)*; Tulane University: New Orleans, LA, USA, 1985.
56. Hoover, W.S. Notes on spatial distribution patterns for three Mexican species of Begonia. *Phytologia* **1979**, *43*, 107–132. [CrossRef]
57. Flora of North America Editorial Committee. *Flora of North America-Magnoliophyta: Magnoliidae and Hamamelidae*; Oxford University Press: New York, NY, USA, 1997; Volume 3.
58. Eggli, U. *Crassulaceae (Dickblattgewächse)*; Ulmer: Stuttgart, Germany, 2003; Volume 4.
59. Castroviejo, S.; Aedo, C.; Laínz, M.; Morales, R.; Munoz Garmendia, F.; Nieto Feliner, G.; Paiva, J. *Flora Iberica, Ebenaceae-Saxifragaceae*; Real Jardín Botánico, CSIC: Madrid, Spain, 1997; Volume 5.
60. Lauber, K.; Wagner, G.; Gygax, A. *Flora Helvetica: Illustrierte Flora Der Schweiz: Mit Artbeschreibungen Und Verbreitungskarten von 3200 Wild Wachsenden Farn- Und Blütenpflanzen, Einschliesslich Wichtiger Kulturpflanzen*; Haupt Verlag: Bern, Switzerland, 2018.
61. CABI; Quiroz, D. Macaranga Tanarius. Available online: https://www.cabi.org/isc/datasheet/32763 (accessed on 9 October 2020).
62. Wu, Z.; Raven, P.H.; Hong, D. *Flora of China: Oxalidaceae through Aceraceae*; Science Press: Beijing, China, 2008; Volume 11.
63. Flora of North America Editorial Committee. *Flora of North America-Magnoliophyta: Vitaceae to Garryaceae*; Oxford University Press: New York, NY, USA, 2016; Volume 12.
64. Welzen, P.C. Revisions and phylogenies of Malesian *Euphorbiaceae*: Subtribe *Lasiococcinae* (Homonoia, Lasiococca, Spathiostemon) and clonostylis, ricinus, and wetria. *Blumea Biodivers. Evol. Biogeogr. Plants* **1998**, *43*, 131–164.
65. Kundu, B.C.; Guha, S. The genus perichasma (*Menispermaceae*). *Adansonia* **1977**, *17*, 221–234.

66. Aiton, W. Oxalis. In *A General History of the Dichlamydeous Plants*; Gilbert & Rivington: London, UK, 1831; Volume 1, pp. 753–768.
67. Groom, Q. Typification of Oxalisbowiei WT Aiton Ex, G. Don (*Oxalidaceae*). *PhytoKeys* **2019**, *119*, 23. [CrossRef] [PubMed]
68. Porter-Utley, K. A Revision of Passiflora, L. Subgenus Decaloba (DC.) Rchb. Supersection Cieca (Medik.) JM MacDougal & Feuillet (*Passifloraceae*). *PhytoKeys* **2014**, *43*, 1.
69. *Flora of North America Editorial Committee Flora of North America-Magnoliophyta: Paeoniacaea to Ericaceae*; Oxford University Press: New York, NY, USA, 2009; Volume 8.
70. Wu, Z.; Raven, P.H. *Flora of China: Brassicaceae through Saxifragaceae*; Science Press: Beijing, China; Missouri Botanical Garden Press: St. Louis, MO, USA, 2001; Volume 8.
71. Grau, A. ; *Mashua, Tropaeolum Tubrosum Ruíz & Pav*; International Potato Center: Lima, Peru, 2003; Volume 25.
72. Hodge, W.H. Three native tuber foods of the high andes. *Econ. Bot.* **1951**, *5*, 185–201. [CrossRef]
73. National Research Council. *Lost Crops of the Incas: Little-Known Plants of the Andes with Promise for Worldwide Cultivation*; National Academy Press: Washington, DC, USA, 1989.
74. Berg, C.C.; Rosselli, P.F.; Davidson, D.W. Cecropia. *Flora Neotropica* **2005**, *94*, 1–230.
75. Radcliffe-Smith, A. Pilea Peperomioides: *Urticaceae. Kew Mag.* **1984**, *1*, 14–19. [CrossRef]
76. Wu, Z.; Raven, P.H.; Hong, D. *Flora of China: Ulmaceae through Basellaceae*; Science Press: Beijing, China; Missouri Botanical Garden Press: St. Louis, MO, USA, 2003; Volume 5.
77. Ernandes, P.; Marchiori, S. The rare water fern marsilea strigosa willd: Morphological and anatomical observations concerning a small population in a mediterranean temporary pond in puglia. *Plant Biosyst. Int. J. Deal. Asp. Plant Biol.* **2012**, *146*, 131–136. [CrossRef]
78. Samain, M.-S.; Vanderschaeve, L.; Chaerle, P.; Goetghebeur, P.; Neinhuis, C.; Wanke, S. Is morphology telling the truth about the evolution of the species rich genus peperomia (*Piperaceae*)? *Plant Syst. Evol.* **2009**, *278*, 1. [CrossRef]
79. Sack, L.; Scoffoni, C. Leaf venation: Structure, function, development, evolution, ecology and applications in the past, present and future. *New Phytol.* **2013**, *198*, 983–1000. [CrossRef] [PubMed]
80. Tadrist, L.; Saudreau, M.; de Langre, E. Wind and gravity mechanical effects on leaf inclination angles. *J. Theor. Biol.* **2014**, *341*, 9–16. [CrossRef] [PubMed]
81. Niklas, K.J.; Paolillo, D.J., Jr. The role of the epidermis as a stiffening agent in tulipa (*Liliaceae*) stems. *Am. J. Bot.* **1997**, *84*, 735–744. [CrossRef]
82. Moonlight, P.W.; Ardi, W.H.; Padilla, L.A.; Chung, K.-F.; Fuller, D.; Girmansyah, D.; Hollands, R.; Jara-Muñoz, A.; Kiew, R.; Leong, W.-C. Dividing and conquering the fastest–growing genus: Towards a natural sectional classification of the mega–diverse genus begonia (*Begoniaceae*). *Taxon* **2018**, *67*, 267–323. [CrossRef]
83. Kaul, R.B. Anatomical observations on floating leaves. *Aquat. Bot.* **1976**, *2*, 215–234. [CrossRef]

 *biomimetics*

 MDPI

Essay

The Challenges of Inferring Organic Function from Structure and Its Emulation in Biomechanics and Biomimetics

Karl J. Niklas [1,*] and **Ian D. Walker** [2]

[1] School of Integrative Plant Science, Cornell University, Ithaca, NY 14853, USA
[2] Department of Electrical and Computer Engineering, Clemson University, Clemson, SC 29634, USA; iwalker@g.clemson.edu
* Correspondence: kjn2@cornell.edu

Abstract: The discipline called biomimetics attempts to create synthetic systems that model the behavior and functions of biological systems. At a very basic level, this approach incorporates a philosophy grounded in modeling either the behavior or properties of organic systems based on inferences of structure–function relationships. This approach has achieved extraordinary scientific accomplishments, both in fabricating new materials and structures. However, it is also prone to misstep because (1) many organic structures are multifunctional that have reconciled conflicting individual functional demands (rather than maximize the performance of any one task) over evolutionary time, and (2) some structures are ancillary or entirely superfluous to the functions their associated systems perform. The important point is that we must typically infer function from structure, and that is not always easy to do even when behavioral characteristics are available (e.g., the delivery of venom by the fangs of a snake, or cytoplasmic toxins by the leaf hairs of the stinging nettle). Here, we discuss both of these potential pitfalls by comparing and contrasting how engineered and organic systems are operationally analyzed. We also address the challenges that emerge when an organic system is modeled and suggest a few methods to evaluate the validity of models in general.

Keywords: adaptation; engineering theory; evolution; form-function; modelling; plants

Citation: Niklas, K.J.; Walker, I.D. The Challenges of Inferring Organic Function from Structure and Its Emulation in Biomechanics and Biomimetics. *Biomimetics* **2021**, *6*, 21. https://doi.org/10.3390/biomimetics6010021

Academic Editors: Olga Speck and Thomas Speck

Received: 8 January 2021
Accepted: 16 March 2021
Published: 18 March 2021

Publisher's Note: MDPI stays neutral with regard to jurisdictional claims in published maps and institutional affiliations.

1. Introduction

Attempts to model or emulate organic structure–function relationships often rely on the fact that the physiological and structural requirements for growth, survival, and reproductive success are remarkably similar for the majority of extant as well as extinct organisms regardless of their phyletic affiliation, even at the molecular level [1–5]. In addition, most of these requirements can be quantified by means of comparatively simple mathematical relationships drawn directly from the physical and engineering sciences [6–11]. Owing in part to the advent and rapid expansion of computational and fabrication technologies, the number of structure–function models has burgeoned in the last few decades to encompass every level of biological organization ranging from molecular self-assembly (e.g., in vitro spindle apparatus self-assembly) to ecological and evolutionary dynamics (e.g., diversification patterns resulting from differential extinction–origination models) (e.g., [2,10,12,13]).

The interest in modeling organic structure–function relationships has expanded perhaps most aggressively in the field of biomechanics and subsequently into biomimetics—an interdisciplinary discipline in which researchers apply the principles of physics, engineering, chemistry, mathematics, and biology to create synthetic inorganic systems or materials that mimic the functions of biological systems, materials, or processes [3,4,14,15]. This approach has resulted in significant progress in both its theoretical and practical applications. This success emerges in part because biological systems have passed the gauntlet of evolutionary time and successive episodes of natural selection, thereby winnowing out failures. In this sense, almost every organic system can be viewed as the result of a series of evolutionary 'experiments' whose continued success reflects the consequences of intense scrutiny [9].

However, extracting insights from these 'experiments' and mimicking the attributes of organic systems presents a number of challenges not least of which is the assumption that structure–function relationships can be inferred correctly and unambiguously.

In addition, by their very nature (as well as their specific epithets), the fields of *bio*mechanics and *bio*mimetics rely heavily on the principles and theoretical insights gained from the application of engineering and physics to analyze biological system in terms that differ in many important ways from those that emerge directly from the study of biological systems. Indeed, as the level of biological enquiry increases from the molecular to the organismic and from the organismic to the level of evolutionary processes, the 'gap' between the physical and biological sciences increases non-linearly and not without substantial practical and theoretical consequences and challenges.

One of the goals of this paper is to compare and contrast the philosophical differences emerging from the physical sciences (specifically engineering and physics) and the 'synthetic' fields of enquiry called biomechanics and biomimetics. The applicability of the fundamental principles of the physical sciences to understand biology is beyond question. No organism can obviate these principles. However, it must be acknowledged that the manner in which these principles describe or predict the behavior of biological systems differs from how they apply to inorganic systems.

Consider for example the concept of entropy. Although it has two very formal definitions (one in the context of thermodynamics and another in terms of statistical mechanics), entropy can be viewed as a measure of a system's disorder, or randomness. In simple terms, the basic principle associated with entropy (the Second Law of Thermodynamics) states that in the absence of any externally applied source of energy, a system will naturally progress from an ordered to a disordered state. Thus, for any process to occur spontaneously, it is necessary that the entropy in the system in which the process occurs decrease. Living things appear to be negentropic—that is, they appear to become more ordered and less random. This seems to be the case when we look at a living cell or a multicellular organism as it grows in size and continues to function normally. Yet, when we consider that every living organism is an open system—that it exchanges matter and energy with its immediate environment—we see that this negentropic perception is incorrect. All living organisms increase the entropy of their environment in two ways—they convert larger organic molecules (e.g., proteins, carbohydrates, and fats) into smaller inorganic molecules (e.g., CO_2, water, NH_3), and they produce heat. The decrease in entropy associated with the growth of an organism is greatly outweighed by the increase in entropy in the organism's environment. A simple example of this is the so called "10% rule", which governs the flow of biomass between two adjacent trophic levels in an ecosystem. Each pound of living "stuff" that is made in the next higher trophic level requires the metabolic conversion of 10 pounds of living matter (into CO_2, water, and heat) from the lower trophic level. Thus, only 10% of the material produced in the lower trophic level appears as biomass in the higher level.

In this paper, some emphasis will be placed on the implications emerging from the use of models in biomechanics and biomimetics. In this context, a model is defined as any physical or conceptual representation of a naturally occurring system or process intended to provide mechanistic insights into the operation of the phenomenon under investigation. As will be seen, this definition distinguishes a model from a prototype or pattern-architype (wherein the functionality of the prototype or pattern-architype is irrelevant to how the prototype or pattern-architype is used). The benefits of modeling physical or biological systems are widely acknowledged across every scientific discipline. This general applicability results from the fact that many systems are unavailable for direct observation or experimentation because they are too big, small, expensive, rare, or fragile to observe or manipulate directly. Modeling such systems to analyze their behavior is a convenient approach to coping with these limitations. Likewise, some systems that are subject to direct observation may be so complex or imperfectly known that modeling provides the only tractable method to explore the details of their operation. Here, models can be used to generate hypotheses that can be empirically examined subsequently. Finally, models are an important part of our pedagogic armamentarium. They provide a method to communicate and summarize complex ideas or operations easily yet accurately. However, it must be acknowledged that as an abstraction (or perhaps more accurately a redaction) of reality, models can obtain fallacious results. Perhaps worse, a model can give the correct answers but for the wrong reasons, e.g., the astrolabe of Claudius Ptolemy (~100–170 CE) yielded correct predictions about lunar phenomena, but nevertheless assumed the Earth was the center of the universe. This potential pitfall is important when the objective of a model is to elucidate mechanistic explanations for a given phenomenon. It must be freely admitted that a mechanistically incorrect model inspiring an engineer or designer to create a novel and ultimately successful solution to a problem is a useful and valuable model. The "trick", however, is to know whether the model is mechanistically valid or invalid.

In the following sections, we discuss the types of models that are typically used in science, compare and contrast the approaches taken in the physical and biological sciences, and suggest criteria that can be used to assess engineering and biological models in general.

2. An Important Caveat about Behavior

It is important to note that the bulk of what follows focuses on plants rather than animals. Acknowledging this focus is essential because the discussion to be presented largely ignores the important role of "behavior" when inferring function from structure. Animals typically manifest behavior that is immediately obvious (and typically easily translated into human terms), whereas plant "behavior" is generally not immediately obvious because it operates on extended time-scales (i.e., over the course of hours, days, or even years) and primarily at the cellular rather than supracellular level. Consider for example the hypodermic functionality of a snake's fang and the unicellular epidermal hairs (trichomes) on the leaves of the common stinging nettle *Urtica dioica*. Both of these structures are capable of injecting a toxic substance subcutaneously by means of exerting pressure on a containment vesicle (i.e., the venom gland and the cell base of the trichome). The snake delivery system accomplishes this by means of provoked neuromuscular contraction (an act of volition) of a complex multicellular venom gland, whereas the nettle injects its cellular contents passively when its acicular tip is fractured and broken off. Arguably, at many levels, the fang and trichome have very little structural resemblance to a medical hypodermic needle. However, the pressure and fluid flow behaviors of all three are similar upon reflection, but only after observing how all three behave dynamically. In the case of the trichome, this kind of observation would require careful microscopic manipulations (which, in turn, necessitate premeditation on the part of the researcher and some degree of fore-knowledge).

Far more important in the context of this paper is a dictum that emerges from what can be called the "no function in structure" principle, which essentially argues that "function" resides in the "behavior" of a system rather than in the "structure" of a system. According to this principle, structure *implements* function, just as a medical hypodermic needle *injects* a liquid or a fang injects venom, the *behavior* of the needle or the fang is in fact the *function* of the needle and the fang. This principle argues that function can only be achieved by the behavior of a system, which can be redacted by examining how parameters, properties, or some other variables of interest change and not on the basis of examining structure. In the context of dynamic systems, the "no function in structure" principle is relevant to both engineering and biology. However, it misses an important point, viz., evolution and natural selection operate on the manifest, integrated phenotype and the phenotype includes structure as well as behavior. In addition, there are many aspects of plant biology in which structure dictates behavior and in which structure has no behavior (e.g., the material properties, structure, distribution, and conformations of the different cell-types in wood dictate the behavior of wood, and the functionalities of a thorn do not involve any manifest behavior). We shall return to this issue but raise it here to keep the "no function in structure" principle in sight.

3. Types of Models

What constitutes a model and how can a model be tested? This question is not easily answered because models come in many forms, are put to different uses, and must survive the challenge of different kinds of tests to gain credibility. Although there are many models, there are only three general types: iconic models, analogue models, and mathematical models [12]. Iconic models are physical conceptions of reality (e.g., scale models). Analogue models are schematic representations of dynamic processes or operations (e.g., photosynthesis rendered in chemical notation). Mathematical models are the most abstract of the three general types because they express reality in terms of computational operations (e.g., the Nernst equation). Importantly, each type of model can be used as a descriptive, behavioral, or decision-making tool. A descriptive model represents the operational relationship, order, or sequencing among the components of a real system (e.g., a reconstruction of a plant used to show organographic or anatomical relationships). Behavioral models are used to predict the response of a real system to perturbation (e.g., a scale model of plant placed in a wind tunnel to estimate wind-induced drag forces and stresses). Decision-making models are used to identify which among available alternative responses of a system is the most likely or favorable according to a priori criteria (e.g., cladistic algorithms and statistical inference models). Complex models can be constructed using two or more types of model in combination as descriptive, behavioral, and decision-making tools.

Regardless of its type or intended use, however, every model is a conceived image of reality and every modeler, therefore, must deal with the ongoing tension between the ideal (conceived image) and the real (observed data) (see [12]). This does not imply that all useful models are complex. A model's complexity depends as much on the objective of modeling as it does on the structure, object, operation, or process being modeled. When put to some simple use, a model can successfully mimic a structure or process with a minimum number of assumptions or stipulations. For example, to evaluate convective cooling, a cow can be modeled as a cylinder (provided its surface area with respect to volume is scaled properly). Indeed, some very useful models bear little resemblance to the actual appearance of the object or system being represented. But it is always true that a model is only as useful as the extent to which its behavior or operation faithfully accords with that of reality. A cylindrical cow is a useful model in terms of understanding the physics of heat dissipation, but, in terms of locomotion, a cylinder rolls whereas a cow walks.

4. Plant Versus Animal Models

Modeling the relationship between animal structure–function relationships has a long and distinguished history (e.g., [16–21]). In comparison, modeling has only recently been used to explore the quantitative relationships between plant structure and function. Perhaps not coincidentally, efforts in biomimetics founded on plant structure–function relationships [22,23] have tended to lag behind those in animals [17–21,24]. While this is likely due in part by a desire to emulate various types of mobility—significantly hindered by the sessile nature of plants—we speculate that the relatively recent emergence of structure–function plant models has contributed to this lag. Nonetheless, within the last few decades, virtually every level of biological organization has been explored for plants, ranging from the level of molecular self-assembly to community structure, ecological interactions, and evolutionary dynamics.

Although the reasons for this burgeoning interest in modeling and mimicking plant structure–function relationships are obvious, the reasons that plant models are latecomers are not obvious, especially since modeling how organic structure and function interrelate is arguably easier for plants than for animals. Unlike animals, the vast majority of plant species perform the same tasks to grow, survive, and reproduce—plants use sunlight to manufacture their living substance, exchange atmospheric gases with their external environment, absorb and transport water and minerals, cope with the mechanical stresses induced by externally applied mechanical forces, and they capture or disperse spores, seeds, fruits, or other similar structures to reproduce or colonize new sites. Thus, unlike the case for animals, the metabolic operations and requirements of plants, both past and present, are nearly identical in their broad outline. Likewise, most plants are sedentary and all lack neurological and muscular systems, and thus they grow, reproduce, and die in much the same location in which they began their existence.

Therefore, plants can be modeled more as "structures" than as "mechanisms" (e.g., [5]), although plants manifest dynamic mechanical behaviors (e.g., [14]). For these reasons, the botanist can turn directly to the techniques and concepts formulated by the physical and engineering sciences to model virtually any plant function requiring energy or mass transport (e.g., photosynthesis, respiration, water and mineral absorption) or involving solid and fluid mechanics (e.g., heat dissipation, mechanical stability, passive or active dispersal of spores or propagules, and the bulk transport or movement of water or air). Analytical geometry can be used to evaluate the capacity of a plant to intercept sunlight; equations such as Fick's law can be used to model the passive diffusion of carbon dioxide, oxygen, or any other substance through its tissues; the transport of water through plant vascular systems can be evaluated by means of the Hagen–Poiseuille equation; and the long-distance passive dispersal of reproductive structures by wind or water can be treated with the aid of Stokes' law or comparatively simple physical or mathematical models. On a macroscopic scale, the ability of any plant to cope with the effects of gravity or drag induced bending moments and stresses can be evaluated with the aid of standard engineering techniques and concepts, whereas ecological interactions among conspecifics or among members of different plant species can be modeled often on the basis of simple allometric "rules" that appear to hold true through across the full spectrum of plant size and all clades.

The botanist is also at advantage because the spectrum of plant body-plans is narrower than that of animals [25]. Whereas the body-plans of the algae are extremely diverse (e.g., unicellular, colonial, filamentous, thalloid, and treelike), the land plant species share the same basic body-plan predicated on one or more cylindrical organs (stems or stem-like analogues) bearing flattened organs (leaves or leaf-like analogues) anchored to a substrate by a variety of means (e.g., rhizoids and multicellular scales) of which roots are the most familiar. Regardless of their apparent complexity, all plant body-plans can be modeled using one or more simple geometric forms (e.g., terete cylinders and oblate or prolate spheroids) whose shape and size can be adjusted independently or scaled with respect to each other.

Modeling any form–functional relationship, however, has its pitfalls. Whereas the fundamental tenet of modeling is that the first principles of physics, chemistry, engineering, and mathematics cannot be violated, it is also true that organisms typically obviate (or at least obscure) the effects of some of these well-known principles by virtue of growth, reproduction, or unique adaptations, aspects of which are sometimes poorly preserved or entirely obscured in the fossil record. For example, Fick's law for passive diffusion always holds true, just as Euler's buckling formula accurately predicts how tall a tree can grow before it elastically deflects under its own weight. But Fick's law is irrelevant for species that actively transport carbon dioxide in the form of bicarbonates (as do some aquatic plants), whereas Euler's buckling formula gives different estimates of the tree height for woods differing in their lignin content and thus stiffness (and assumes that mechanical constraints rather than hydraulic constraints limit tree height). Even the most credulous may reasonably doubt that organisms have properties that make the direct application of closed-form engineering solutions to understand their biology highly problematic.

5. Comparing Standard Practices (and a Case Study)

Physical principles can be used to interpret and understand biological phenomena, but it must be acknowledged that the standard practices used in the physical and biological sciences differ in many important ways (Table 1). Consider how an engineer approaches a challenge. In most cases, the working environment and design specifications for a machine, structure, or material are specified a priori. There are exceptions naturally. The design and construction of machines for deep-space or other-worldly exploration require considerably more thought. However, in general, there is no second guessing under most situations. In addition, the structure and the materials used to construct the required artifact can be altered during the design and implementation process. There is no historical constraint on design or construction practices. Finally, it is generally true (albeit not invariably so), that an engineered artifact has one or only a comparatively few functional obligations—a toaster, hair drier, phonograph, or comb has but one primary function. It must be recognized that a toaster or hair drier can be used to heat a room, a phonograph can be used to hypnotize an unwilling musical student, and a comb can be used as a musical instrument. Consequently, the primary functional obligations of each of these engineered artifacts have nothing to do with other potential functionalities. In contrast, a foliage leaf must perform multiple functions simultaneously, as for example light-interception, gas exchange, mechanical stability, defense against herbivory or pathogen attack, hydraulic requirements, heat dissipation, and a number of other developmental "obligations" attending its maturation from a small mass of primordial cells emerging from an apical meristem. In passing, however, it is useful to note that "unintended" functionalities (e.g., the use of a comb as a musical instrument or a hair drier to defrost a windowpane) provide evolutionary opportunities. Leaves probably evolved primarily as photosynthetic structures, but over the course of organic evolution they have become modified to yield spines, bracts, sepals, petals, stamens, carpels, and digestive organs.

Table 1. Comparisons between the standard practices of engineering vs. biomechanics.

Engineering	Biomechanics
1. Working environment specified a priori	1. The environment is variable
2. Design specifications are known	2. The environment has to be examined
3. Function is specified a priori	3. Function is inferred ad hoc
4. Structure and materials can be altered	4. Structure and materials have historical legacies
5. The structure typically has one function (thus, the function can be maximized)	5. The structure typically has multiple functions (thus, functions are reconciled)
6. Often, only a comparatively few accurate measurements are required to determine structural or material properties *	6. Many measurements are required because of natural biological variations in structure or materials

* The accurate measurement of the mass of an electron can be used to estimate the mass of all electrons in the universe, whereas the accurate measurement of the mass of a gerbil provides the mass of a single animal at a particular time in its life.

In contrast, each of the engineering "standard practices" cannot be applied when dealing with biological structure–function relationships. Most organisms exist in a highly variable environment (e.g., air- or water-flow can vary many orders of magnitude, even in a few minutes), thereby requiring the biomechanicist to examine the "work-space" empirically. Perhaps worse, function has to be inferred. In some cases, this is not a huge challenge. The functionality of a snake's hypodermic-like fangs seemingly requires little interpretation. As noted in the introduction, the behavior of a snake's fang is sufficient to infer its function, arguably even in the absence of understanding it structure. Indeed, one can argue that "structure" cannot be used to infer "function" because "structure" implements function, whereas behavior is "function". This is a sagacious dictum emerging from the "no function in structure principle" that reappears in various guises in both engineering theory and practice. This principle argues that function resides in the implementation of function, i.e., function resides in the behavior of a system. However, in many cases, biological "behavior" may be entirely absent. For example, what is the function of a rose's prickle (the structure that is incorrectly called a "thorn" on a rose stem is actually a prickle and not a thorn)? Many believe that it functions as an herbivore deterrent. Yet, it can also function to dissipate heat, or as a grappling hook (indeed, prickles seem to be most developed in cultivars of climbing roses). Arguably, the rose prickle need not have any biological function whatsoever to provide a model for a better grappling hook or thermal vent. In this sense, a prickle can serve as a "model" for designing a device that has nothing to do with its biological functionality. However, such a model provides no mechanistic insights into the evolution and biology of a rose's "thorn". Consequently, in the context of this review, such a model is not a model—it is a prototype or pattern-architype.

Other problems present themselves to the biomechanicist. Structural configurations and tissue (material) compositions are constrained by evolutionary and developmental features that cannot be changed easily. As mentioned earlier, many biological structures also have multiple functions. Thus, although an engineer can try to maximize the performance of a machine performing one function, the biologist must interpret structures that most likely evolved to reconcile contrasting functional obligations.

It cannot escape attention that there is also a rather important contrast between physics and biophysics (Table 1). In most instances, one accurate measurement of a fundamental process or object (e.g., electrical resistance or the mass of an electron) with reliable equipment is sufficient to predict the behavior of the same process or object universally. A photon of a particular wavelength will behave the same way universality. In contrast, the accurate and precise measurement of the weight of a gerbil only reveals the weight of a very particular animal on a very particular day.

The foregoing may seem overly simplified and somewhat abstract. Certainly, all the limitations outlined in Table 1 can be overcome by the biomechanicist with sufficient insight and effort. That is not the real issue. What is important is dealing with the assumptions emerging from applying engineering theory to a biological problem. In order to illustrate this concern, we turn to a very simple case study—the application of engineering pertaining to the construction of a vertical support member to understanding the functionality of a tree trunk (Table 2).

With very few exceptions, the equations used to calculate the maximum load a columnar support member assume that the member has a uniform, unflawed geometry and that it is composed of a more or less uniform material. These equations also assume that the support member is subjected to a uniform loading condition (or that it experiences a principal stress application) under uniform working conditions. Most of these assumptions are implicit in the mathematics underlying the classic Euler–Greenhill formula used by many to estimate maximum tree height:

$$H_{crit} = C \left(\frac{E}{\rho g} \right)^{1/3} D^{2/3},$$

where H_{crit} is the theoretical critical buckling height of a column, C is a normalization constant that describes the uniform taper of the column (assumptions 1 and 2; see Table 2), E is Young's modulus (assumption 3), ρ is the density of the construction material (assumption 3), g is the acceleration of gravity (assumptions 4 and 5), and D is the basal diameter of the column (assumptions 5 and 6). Curiously, the Euler–Greenhill formula makes another (often neglected) assumption—it assumes that the maximum height of a column is limited for (static) mechanical reasons rather than some other limiting factor. In the case of a tree, the maximum height is often limited by the rate at which water can be elevated to the upper reaches of the canopy at a rate sufficient to sustain stomatal conductance (and thus sustain the growth of first-year shoots).

Table 2. Six assumptions (and qualifications) made when engineering a columnar support member.

○   Assumption 1: Specified Geometry (no Growth Variations)

Qualification: *This is typically, but not invariably true. "Growth" (e.g., extension via controlled sliding of—themselves non-growing—elements) is almost always prespecified, and usually part of the design specifications for desired functionality.*

○   Assumption 2: No production flaws (no fractures, or "knots")

Qualification: *None. This assumption is almost universally made.*

○   Assumption 3: Uniform (more or less) material composition

Qualification: *This assumption is almost always made, though more recent work in composite materials necessitates a relaxation of this assumption.*

○   Assumption 4: Uniform stress application (compression, tension, or torsion)

Qualification: *Typically, one stress application (the extreme case) dictates the design, e.g., gravity or some uniform field applied to the whole member(s), or a small, finite number of applied point forces/moments, such as a point contact with the environment during drilling.*

○   Assumption 5: Below yield stress conditions (designed not to break)

Qualification: *This assumption is typically tested* ad hoc *in physical prototypes* via *destructive testing.*

○   Assumption 6: Uniform workplace conditions (controlled climate)

Qualification: *This is true for most engineering applications, but with some spectacular exceptions. For example, in Space systems in orbit, temperatures change over ranges of many hundreds of degrees in fairly short periods of time, and those systems have to be designed with those constraints in mind.*

6. Technological Challenges in Biomimetic Engineering

From the engineering perspective, once provided with underlying principles and understanding established in the modeling of organic systems, a further challenge emerges: to reproduce their functionality given available technologies. Although we live during a time in which technology has dramatically altered the human condition, our "age of technology" is infinitesimal in the context of evolution's "engineering" of organic systems. Human-engineered materials and processes are, correspondingly, often quite crude and primitive when compared with organic systems with similar functionality. In many cases, even when we understand and can suitably model the structure and function of organic systems, the complexity inherent in their operation is beyond the capability of currently available technology to emulate [3].

As an example, efforts to recreate the biologically well-understood "muscular hydro-static" structure of cephalopod arms and similar organs [26] in robotic variants in the early 2000s (e.g., [27,28]) were significantly limited by limitations in the functionality of artificial muscles. Mammalian muscles, which form the core structure of cephalopod arms (as well as elephant trunks and human tongues, among other familiar structures) are capable of strains well beyond that of their artificial equivalents. When combined with inherent limitations in the scalability, strength, and power generation of artificial muscle materials, the robotic versions of cephalopod arms produced lacked the complexity and hence several of the key movements (for example significant extension combined with torsion) that make their natural counterparts so effective (Figure 1).

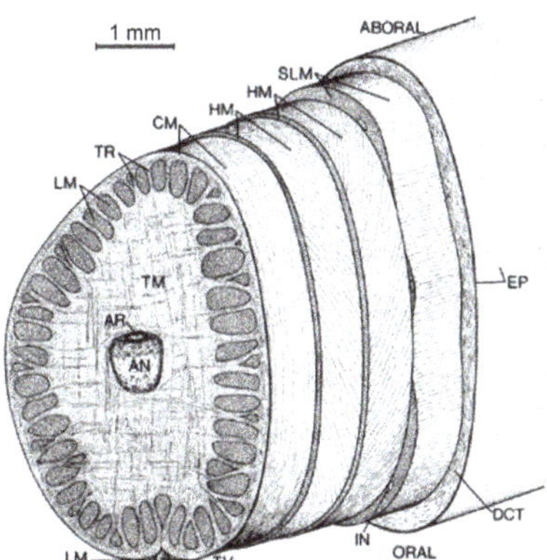

Figure 1. (Top): cross section of cephalopod arm musculature [29]; (**bottom**): robotic "octopus arm" [27].

Similar limitations have confronted engineers attempting to synthesize robotic versions of plants which can mimic the growth of structures such as roots and stems, in order to explore and examine imprecisely known environments. Some approaches have deployed thin telescoping tube structures [14]. However, "growing" such structures requires movement, relative to the environment, of large portions of the structure—the translating tube(s)—which in turn creates significant frictional resistance along the backbone when in contact with the environment. Plants avoid such issues by growing at the tips of their structures. Thus, environmental friction arising from plant growth is reduced to that at the tip. Experiments to mimic and exploit this behavior by depositing (3D printing) material directly at the tip of "robot roots" have been conducted [30]. However, the size of the technology currently required to 3D print suitable materials drastically limits the scalability of the resulting structures, with diameters at or below the size of the human finger not feasible at the present time.

It is worth noting that even when the structure/function relationships of organic systems are well understood, directly mimicking the organic structure may not meet the design goals of the engineer. Consider the oft-quoted argument against biomimetics, that if aviation engineers had restricted their efforts to reproducing the flight of birds, we would not today be flying at high speeds in airplanes. In the development of fixed-wing aircraft, the underlying principles of powered flight (notably the aerofoil wing profile), rather than the organic structures demonstrating it, proved the key to successful engineered systems. However, there are many situations in which not only the function but also the form of organic structures needs to be replicated via engineering (e.g., in prosthetics), and others in which either the underlying principles are poorly understood, and for which biomimetic engineering can be used to explore those principles. In this work, we focus on such situations, and in the following discuss in more detail inherent difficulties arising from mismatches in assumptions in standard practices of engineering, biomechanics, and physics in contrast to the environmental and evolutionary constraints arising in biophysics.

7. Criteria and Challenges for Assessing a Model

As noted, the use of modeling is generally unavoidable in almost every scientific and analytical discipline. Consequently, it is necessary to evaluate their performance empirically, which requires explicit metrics with which to measure the concurrence of a model's predictions and the empirical data. In many cases, the metrics are de minimis statistical (e.g., 95% confidence levels) (Table 3). Under some circumstances, other metrics are necessitated (e.g., measuring mechanical or robotic performance and similitude to a biological prototype). In this context, it is essential to specify the scope of a model, thereby drawing a sharp distinction between prediction and extrapolation. By its very nature, the scientific enterprise typically attempts to predict the behavior of a system beyond the scope of the data used to model a system (Table 3). However, attempts to extrapolate a behavior or to draw conclusions beyond the extent of empirical observations or existing statistical trends are confronted with the danger that the methodological approaches or the range of behavior or trends will not continue beyond the scope of a model, i.e., interpolation within the scope of a model, differs profoundly from extrapolation beyond the scope of a model. Yet, another important test of a model is whether properties or behaviors emerge from the model that are not explicitly or directly engrained within the model's structure. These criteria and challenges to validating a model surface across almost all biomechanical and biomimetic studies because most models are based on three generic assumptions: (1) the assumption that a single process is responsible for the pattern or behavior observed (either in the data or a physical manifestation of the system to be modeled), (2) the related assumption that one major process dominates an observed pattern or behavior, and (3) the assumption that biotic interactions are correctly interpreted and easily dissected. We have already noted that in some cases even an inaccurate model can be useful when it can be used by engineers or designers to create a useful result. For example, the model used in [31] to analyze observed circumnutations in plants, while not directly reflecting plant growth, has been used to effectively model the extension capability of spring-loaded vine-inspired robots [14]. However, as noted previously, the word "model" in this case refers to a "prototype" and differs substantively from a "model" intended to reveal mechanistic insights into the entity being modeled.

Table 3. Three criteria for evaluating models and the challenges they present.

Criteria	Challenges
1. Fit simulation to empirical data (e.g., correlation and cross-validation analyses)	1. Specify metrics to measure concurrence (e.g., specify a—values and r^2—values)
2. Specify the model's scope (e.g., interpolation analyses)	2. Draw a sharp distinction between prediction and extrapolation
3. Identify emergent properties	3. Avoid the fallacy of causation

Setting aside the scientific challenges that face modeling a biological system, in our combined experiences there exists a sociological challenge. It seems clear to us that, in any non-trivial situation for which a biomimetic or biomechanical approach is motivated, it is essential to actively engage and combine the expertise of engineers and biologists drawn from the relevant specialties of each domain. One obstacle to progress in such intended synergies can be "culture shock" across domains. Engineers, when first confronted with the details of organic systems, tend to find the complexity of such systems quite intimidating. They can find the challenge of interpreting the available knowledge to extract underlying principles relevant to their engineering options overwhelming and rather demotivating. Experts in biology can be disappointed that the engineering options are often primitive and limited compared to organic structure and processes. Often, particularly in early prototypes, the engineered analogues offer somewhat sterile abstractions of the organic systems with which biologists are familiar. Enthusiasm and continued engagement requires open-mindedness, persistence, and good communication between all parties, which can sometimes be a challenge for people conditioned to specialized ways of thinking.

Author Contributions: Conceptualization: K.J.N.; writing: K.J.N. and I.D.W. Both authors have read and agreed to the published version of the manuscript.

Funding: This research was funded by U.S. National Science Foundation, grant number IIS-1718075.

Institutional Review Board Statement: Not applicable.

Informed Consent Statement: Not applicable.

Data Availability Statement: Not applicable.

Conflicts of Interest: The authors declare no conflict of interest.

References

1. Sarikaya, M.; Tamerler, C.; Jen, A.K.Y.; Schulten, K.; Baneyx, F. Molecular biomimetics: Nanotechnology through biology. *Nat. Mater.* **2003**, *2*, 577–585. [CrossRef]
2. Tang, Z.; Wang, Y.; Podsiadlo, P.; Kotov, N.A. Biomedical applications of layer-by-layer assembly: From biomimetics to tissue engineering. *Adv. Mater.* **2006**, *18*, 3203–3224. [CrossRef]
3. Vincent, J.F.V.; Bogatyreva, O.A.; Bogatyrev, N.R.; Bowyer, A.; Pahl, A.K. Biomimetics: Its practice and theory. *J. R. Soc. Interface* **2006**, *3*, 471–482. [CrossRef]
4. Bhushan, B. Biomimetics: Lessons from nature—An overview. *Philos. Trans. R. Soc. A Math. Phys. Eng. Sci.* **2009**, *367*, 1445–1486. [CrossRef]
5. Koch, K.; Bhushan, B.; Barthlott, W. Multifunctional surface structures of plants: An inspiration for biomimetics. *Prog. Mater. Sci.* **2009**, *54*, 137–178. [CrossRef]
6. Thompson, D.A.W. *On Growth and Form*; Cambridge University Press: Cambridge, UK, 1942.
7. Raup, D.M.; Michelson, A. Theoretical morphology of the coiled shell. *Science* **1965**, *147*, 1294–1295. [CrossRef]
8. Niklas, K.J. *Plant Biomechanics: An Engineering Approach to Plant Form and Function*; University of Chicago Press: Chicago, IL, UK, 1992.
9. Niklas, K.J. *The Evolutionary Biology of Plants*; University of Chicago Press: Chicago, IL, UK, 1997.
10. Vincent, J.F.V. Biomimetic modelling. *Philos. Trans. R. Soc. Lond. B* **2003**, *358*, 1597–1603. [CrossRef] [PubMed]
11. Vincent, J.F.V. Biomimetics–A review. *Proc. Inst. Mech. Eng.* **2009**, *223*, 919–939. [CrossRef]
12. Schwartz, R. *Biological Modeling and Simulation*; The MIT Press: Cambridge, MA, USA, 2008.
13. Wanieck, K.; Fayemi, P.E.; Maranzana, N.; Zollfrank, C.; Jacobs, S. Biomimetics and its tools. *Bioinspired Biomim. Nanobiomaterials* **2017**, *6*, 53–66. [CrossRef]
14. Wooten, W.; Walker, I.D. Vine-inspired continuum tendril robots and circumnutations. *Robotics* **2018**, *7*, 58. [CrossRef]
15. Iida, F.; Ijspeert, A.J. Biologically inspired robotics. In *Springer Handbook of Robotics*; Siciliano, B., Khatib, O., Eds.; Springer: Berlin, Germany, 2016; pp. 2015–2034.
16. Niklas, K.J. Modeling fossil plant form-function relationships: A critique. *Paleobiology* **2000**, *26* (Suppl. 4), 289–304. [CrossRef]
17. Richter, R. Psychische Reaktionen fossiler Tiere. *Palaeobiologica* **1928**, *1*, 226–244.
18. Ostrom, J.H. A functional analysis of the jaw mechanics in the dinosaur Triceratops. *Postilla (Peabody Mus. Nat. Hist. Yale Univ.)* **1964**, *88*, 1–35.
19. Rudwick, M.J.S. The inference of function from structure in fossils. *Br. J. Philos. Sci.* **1964**, *15*, 27–40. [CrossRef]
20. Clarkson, E.N.K. Schizochroal eyes and vision in some Silurian acastid trilobites. *Palaeontology* **1966**, *9*, 1–29.
21. Stanley, S.M. *Relation of Shell Form to Life Habits in the Bivalvia (Mollusca)*; Geological Society of America: Boulder, CO, UK, 1970; Volume 125.
22. Martone, P.T.; Boller, M.; Burgert, I.; Dumais, J.; Edwards, J.; Mach, K.; Rowe, N.; Rueggeberg, M.; Seidel, R.; Speck, T. Mechanics without muscle: Biomechanical inspiration from the plant world. *Integr. Comp. Biol.* **2010**, *50*, 888–907. [CrossRef]
23. Mazzolai, B.; Beccai, L.; Mattoli, V. Plants as model in biomimetics and biorobotics: New perspectives. *Front. Bioeng. Biotechnol.* **2014**, *2*, 1–5. [CrossRef]
24. Fish, F.; Kocak, D.M. Biomimetics and marine technology: An introduction. *Mar. Technol. Soc. J.* **2011**, *45*, 8–13. [CrossRef]
25. Niklas, K.J. The evolution of plant body plans—A biomechanical perspective. *Ann. Bot.* **2000**, *85*, 411–438. [CrossRef]
26. Kier, W.M.; Smith, K.K. Tongues, tentacles and trunks: The biomechanics of movement in muscular-hydrostats. *Zoo. J. Linn. Soc.* **1985**, *83*, 307–324. [CrossRef]
27. Walker, I.D.; Dawson, D.; Flash, T.; Grasso, F.; Hanlon, R.; Hochner, B.; Kier, W.M.; Pagano, C.; Rahn, C.D.; Zhang, Q. Continuum robot arms inspired by cephalopods. In Proceedings of the SPIE Conference on Unmanned Ground Vehicle Technology VII, Orlando, FL, USA, 27 May 2005; pp. 303–314.
28. Laschi, C.; Mazzolai, B.; Mattoli, V.; Cianchetti, M.; Dario, P. Design of a biomimetic robotic octopus arm. *Bioinspir. Biomim.* **2009**, *4*, 1–8. [CrossRef]

29. Kier, W.M. The functional morphology of the musculature of squid (Loliginidae) arms and tentacles. *J. Morph.* **1982**, *172*, 179–192. [CrossRef]
30. Sadeghi, A.; Tonazzini, A.; Popova, I.; Mazzolai, B. Robotic mechanism for soil penetration inspired by plant root. In Proceedings of the IEEE International Conference on Robotics and Automation, Karlsruhe, Germany, 6–10 May 2013; pp. 3457–3462.
31. Bastien, R.; Meroz, Y. The kinematics of plant nutation reveals a simple relation between curvature and the orientation of differential growth. *PLoS Comput. Biol.* **2016**, *12*, e1005238. [CrossRef]

MDPI
St. Alban-Anlage 66
4052 Basel
Switzerland
Tel. +41 61 683 77 34
Fax +41 61 302 89 18
www.mdpi.com

Biomimetics Editorial Office
E-mail: biomimetics@mdpi.com
www.mdpi.com/journal/biomimetics